# LEOMA AND THE
# US LASER INDUSTRY

# LEOMA AND THE US LASER INDUSTRY

## The Good and Bad Moves for Trade Associations in Emerging High-Tech Industries

C. BRECK HITZ

IEEE PRESS

WILEY

For general information on our other products and services or for technical support, please contact our
Customer Care Department within the United States at (800) 762-2974, outside the United States at
(317) 572-3993 or fax (317) 572-4002.

Wiley also publishes its books in a variety of electronic formats. Some content that appears in print may
not be available in electronic formats. For more information about Wiley products, visit our web site
at www.wiley.com.

*Library of Congress Cataloging-in-Publication Data is available.*

ISBN: 978-1-118-01024-2

Printed in the United States of America

# CONTENTS

# PREFACE

As its title suggests, this book examines some of the issues that confront a trade association representing a newly developed high-technology industry. In the late 1980s, the US laser industry had sales approaching a billion dollars, but was lacking any form of internal cohesion. Perhaps there will be emerging industries with those characteristics in the second and third decade of the 21st century, and hopefully this book will be helpful to those seeking to guide those industries.

For the U.S. laser industry in 1987, the stimulus was the sudden appearance of several challenges that demanded a coherent response for all the companies involved. The Laser and Electro-Optics Manufacturers' Association (LEOMA) was created, and was mostly successful in addressing those issues. This book describes those issues and how they were addressed, and continues to tell how the lack of subsequent, pressing issues ultimately led to LEOMA's dissolution.

In creating this narrative, it has been advantageous to have worked from the same office—an office with something over 100 ft of bookshelves—for a quarter century. Everything is still here. All the old LEOMA newsletters, all the minutes from LEOMA Board meetings, all the agendas from laser standards committees, all the minutes from conference steering-committee meetings . . . there is even a DOS 1.0 manual on those shelves somewhere.

And it is a good thing those old documents are still available, because the human memory is not a particularly reliable mechanism. In reviewing my notes and memoranda, I discovered that in 1989 I attended a three-day meeting of officials from State, Defense, and Commerce in Albuquerque. The meeting had been precipitated by LEOMA's lobbying for reform of export controls, and apparently it was pivotal to achieving the successful reform described in Chapter 5. I have absolutely no recollection of that meeting. Had I not recently read my own notes from that meeting, I would have denied that it ever took place.

A reviewer of the manuscript for this book complained that the first chapter gives away the ending. Indeed, it does. In fact, I guess I gave away the ending in the second paragraph of this preface. The first chapter gives a bird's-eye overview of LEOMA and its activities. Each of the subsequent chapters examines more closely one of those activities. My hope is that the entire book will serve as an accurate record of this small part of U.S. business history.

C. BRECK HITZ

# 1

# LEOMA AND THE U.S. LASER INDUSTRY

Trade associations are funny things. They bring together companies of different sizes and shapes to address common issues. But unlike political parties or social networks, whose members also address common issues, trade associations bring together entities that not only compete with each other, but also often dislike and even distrust each other. Andrew Procassini, the long time executive director of the Semiconductor Industry Association, titled his book[1] *Competitors in Alliance*, and that is exactly what a trade association is: an uneasy, awkward alliance of often-fierce competitors.

The U.S. laser industry has historically been very competitive. The second major laser company created in the United States, Coherent—or "Coherent Radiation Labs" in those days—was formed in 1966 when Jim Hobart parted ways with the first company, Spectra-Physics, and set up his own shop developing and manufacturing carbon dioxide lasers. The personal animosity between Hobart and one of Spectra-Physics' founders, Herb Dwight, flavored the industry for many years.[2]

---

[1]Procassini, Andrew, *Competitors in Alliance*, Quorum Books, 1995.

[2]The animosity between the companies sometimes bordered on paranoia. In the late 1970s, when I was writing for *Laser Focus* magazine, I had occasion to visit the lab where Spectra-Physics was developing hard-sealed HeNe lasers. The engineer in charge told me the lab was located in the middle of the building because if the lab had windows, the Coherent engineers lurking in the bushes outside would be able to steal the processes Spectra was developing.

---

*LEOMA and the U.S. Laser Industry: The Good and Bad Moves for Trade Associations in Emerging High-Tech Industries*, First Edition. C. Breck Hitz.

Nonetheless, by the early 1980s, several issues were emerging that underlined the need for a trade association among U.S. laser manufacturers. Industry leaders were mulling over the logistics of launching some sort of trade association. A 1983 editorial[3] in *Lasers & Applications* magazine—at that time one of the leading trade publications in the industry—explicitly called for the creation of a trade association, citing several pressing issues.

Highest on the *Lasers & Applications* list was the need to disseminate information about lasers to the nation's manufacturing base. Although lasers could perform many tasks better than conventional tools, manufacturers in general were reluctant to adopt lasers because they were too unknown and unproven. A trade association could be more effective in moving laser techniques into widespread use than a loose and uncoordinated collection of manufacturers, each hawking its own products and often denigrating the products of its competitors.

Export controls, imposed on lasers because they have military as well as civilian applications, were a major hindrance to the growth of international sales in the 1980s. Individual companies lacked the resources required to launch a major revision of those controls, but a trade association, supported by the entire industry, might undertake such a task.

Legal matters and litigation were another important issue. Although the laser was invented in 1960, the U.S. Patent Office issued several basic patents two decades later. Attorneys for Gordon Gould, the inventor who had been awarded the patents, initiated a lawsuit against a small company, General Photonics. Burt Bernard, the president of that company, gave up in despair because he lacked the finances to mount a plausible defense. The lawsuit succeeded and General Photonics went out of business. Armed with that victory, the attorneys took aim at other lasermakers. "A laser trade association might facilitate a more equitable settlement of this dispute," *Lasers & Applications* said, "than the individual skirmishes now taking place."

Many trade associations act as spokesmen for their industries to the U.S. government, and here again a trade association could amplify the voice of the laser industry in matters ranging from safety and education to taxation and regulation.

At about the same time the *Lasers & Applications* editorial was published, one of the industry's professional societies, the Laser Institute of America (LIA), formed a subcommittee, christened the Laser Industry Council (LIC), to address industry concerns. One of the subcommittee's early meetings was held at the California home of Milton Chang (see Figure 1.1), then president of Newport Corporation. Glenn Sherman (see Figure 1.2), who was president of Laser Power Optics and was beholden to Chang for his investment in

---

[3]Hitz, Breck, *Lasers & Applications*, February 1983, p. 20.

**FIGURE 1.1**   Several of the earliest organization meetings that led to the creation of the Laser Association of America, and ultimately LEOMA, took place in Milton Chang's house.

Laser Power, was invited ("summoned" was the word Sherman used, chuckling, as he described events to me recently) to the meeting. When Sherman arrived, he was met by several key LIC players, including Dean Hodges of Newport and Dale Crane of Uniphase, who congratulated him on being the new president of the Laser Industry Council.

**FIGURE 1.2**   Glenn Sherman—shown here at the groundbreaking for his new company, Laser Power Optics—was the LAA's first president.

**FIGURE 1.3**   Hank Gauthier's decision to have Coherent join the LIC was critical in getting the organization started.

"It reminded me of the old joke about the sergeant asking for volunteers," Sherman told me. "Everybody but the new recruit took a step backward."

But Sherman took his new responsibility seriously, recalled Jerry Glen, who was the LIA's technical director at the time. There were many meetings at Chang's house, Glen told me, including one that lasted a whole weekend. One of Sherman's first tasks was to increase the membership beyond the few initial members. He realized that the LIC would never get any traction without the presence of the two industry giants, Spectra-Physics and Coherent. But the animosity between those two companies' leaders—Herb Dwight at Spectra and Jim Hobart at Coherent—seemed to preclude their working together within any organization.

Sherman's solution to this contretemps was to appeal to Coherent's second-in-command, Hank Gauthier (see Figure 1.3). There was no animosity between Gauthier and Dwight, and in fact Dwight had attempted to recruit Gauthier to Spectra years earlier, before Gauthier had joined Coherent. Gauthier liked the idea of an industry association, he recalled recently, because it could enhance the industry in general. "I always thought you had to develop markets first, market share second."

And Gauthier had no problem convincing Hobart, his boss, of the merits of the LIC. "Hobart could care less," he recalled. His boss's entire focus was on the technical development of new products; he wasn't interested in political issues like government regulations and trade associations.

Gauthier's concluding that an industry association could be more effective than individual companies at achieving needed reforms closed the deal, Sherman told me. Coherent joined the LIC, as did Spectra-Physics, and the LIC had achieved critical mass.

But the LIC faced obstacles in addressing industry problems because the LIA's status as a 501(c)(3) nonprofit prohibited lobbying or grassroots political activity, which is precisely what the LIC needed to do. At an LIC meeting in Los Angeles in January 1985, Moe Levitt, the publisher of *Laser Focus* magazine, and Dave Belforte of Belforte Associates argued that the LIC ought to separate from the LIA and become an independent 501(c)(6) nonprofit, whose political activities would not be restricted.

I was on the LIA Board of Trustees in 1985, and I recall a particularly stormy board meeting at the CLEO conference in Baltimore in May of that year. Milton Chang was LIA's president, and he and Hank Gauthier, another LIA board member, were in support of a motion to dissolve the LIC and create a separate entity, a 501(c)(6) nonprofit, in its place. The new corporation would not be hamstrung by the LIA's restriction on lobbying and other activities. It was a contentious issue, and several board members were strongly opposed to the concept because it would diminish LIA's involvement with the laser industry.

The LIA board meeting was simultaneous with the conference reception that evening, and it was deemed politically unwise for the entire board to skip the reception. So in the middle of the LIC debate, the meeting was suspended for an hour so that the participants could put in an appearance at the reception. When the board reconvened, the mood was mellower (wine and beer had been available at the reception) and the opposition to dissolving the LIC had lessened. The board approved the creation of a new entity, the Laser Association of America (LAA), which would apply to the IRS for 501(c)(6) nonprofit status. The LAA's initial officers were Glenn Sherman of Laser Power Optics as president, Ron Kirschner of the Institute of Applied Laser Surgery as secretary, and Kathy Laakmann of Laakmann Electro-Optics as treasurer. There were 16 founding members (see Table 1.1).

In 1986, I was executive editor and a partial owner of *Lasers & Applications* magazine—which by that time had changed its name to *Lasers &*

**TABLE 1.1   Founding Members of the Laser Association of America**

| | | |
|---|---|---|
| Apollo Lasers | Burleigh Instruments | Coherent |
| EG&G | Institute of Applied Laser Surgery | Laakmann Electro-Optics |
| *Laser Focus* | Laser Mechanisms | Laser Power Optics |
| *Lasers & Applications* | Laurin Publishing | Newport |
| Oriel | Quantronix | Spectra-Physics |
| Uniphase | | |

*Optronics*—and my colleagues and I were entertaining offers to sell the magazine. (We completed the sale a year later, and the entire editorial staff departed.) I was teaching my course, *Understanding Laser Technology*, frequently at laser companies, and working on several consulting contracts. But I was intensely aware of the LAA and its activities, and it seemed to me that it was floundering for lack of manpower. The all-volunteer LAA Board of Trustees, composed of people who all had full-time jobs running companies, lacked the time to effectively address all the issues on the table. Glenn Sherman, whose multiyear presidency began with the LIC and continued to the LAA, devoted so much time to that undertaking that his company suffered from his absence, he recalled in a recent conversation.

In August 1986, I phoned Jon Tompkins, an LAA board member whom I knew well from my years at *Lasers & Applications* and *Laser Focus*, and asked what he would think of my serving as part-time, paid staff for LAA. "Very positive," Tompkins responded. And that conversation marked the beginning of an undertaking that would occupy the next 20 years of my professional career.

My previous commitments to teaching and consulting prevented my starting at LAA before June 1987, and even then I had to restrict my involvement to half time. But at its January meeting that year, the LAA unanimously approved my appointment as executive director and raised dues significantly to cover the new expense. For large companies, annual membership went from $1200 to $5000, and for small companies, from $300 to $500, where "large" and "small" were defined as over $20 million in annual sales and under $600,000, respectively. Dues for companies between those extremes were raised similarly.

Probably the most-pressing issue for laser companies in early 1987 was the Gordon Gould laser patents. A decade earlier—but almost two decades after the laser had been invented—the U.S. Patent Office awarded two fundamental laser patents to an inventor named Gordon Gould. During the early years of the laser industry—the 1960s and 1970s—companies had been paying modest royalties to a patent held by Arthur Schawlow and Charles Townes, who had filed their claim in July 1958. Gould filed a claim in April 1959, which had been denied due to the earlier claim by Schawlow and Townes. But Gould pressed his claim, arguing that his notebook entries predated the work by Schawlow and Townes, and in 1977 the Patent Office awarded Gould a patent on optical pumping, one of two primary methods of energizing a laser. In 1979, the Patent Office awarded Gould a second patent, the so-called "use" patent, which covered virtually every use that a laser could be put to. Moreover, Gould had additional patents pending on collisional pumping— the other primary method of energizing lasers—and Brewster windows, a crucial optical element in many lasers.

During the ensuing years, battles raged over the laser patents, with the Patent Office reexamining the original Gould patents and countless appeals launched in courtrooms across the country. The stakes were huge: Gould and his associates demanded much larger royalties than had been paid on the original Schawlow–Townes patent, and laser companies' sales were many times greater than they had been in the early years.

Then, on July 11, 1986, a federal judge directed the Patent Office to issue Gould a patent on collisional pumping, raising the stakes for laser manufacturers even higher. And in November of the same year, the Patent Office Board of Appeals validated Gould's original patent on optical pumping, but rejected the "use" patent. Gould and his associates appealed the rejection, and launched a major effort to enforce the optical pumping patent.

So in January 1987, the members of the Laser Association of America were extremely concerned about the Gould patents. Richard Samuel, the president of Gould's patent-holding company, Patlex, addressed the April 1987 meeting of the LAA board, arguing that the years of legal battling were coming to an end, and advising the companies to accept the fact that they would soon be paying significant royalties to Gould.

There was talk of banding together under LAA to negotiate more favorable terms than could be obtained by individual companies. But the reality was that it was too late. In July 1987, the first major laser manufacturer—Lumonics, a Canadian company—signed an agreement with Gould and his associates to pay royalties on optically pumped lasers sold in the United States.

Another blow landed in August 1987, when the Patent Office announced it would not appeal the earlier court decision to authorize the issuance of the collisional pumping patent. Between the collisional pumping and optical pumping patents, Gould and associates now held patent rights on the vast majority of lasers manufactured in the United States.

Other companies began signing agreements to honor the Gould patents: Kodak, Amdahl, Chrysler, EverReady, and Union Carbide. And in a stunning development in December 1987, a major laser manufacturer—Control Laser of Orlando, Florida—lost the patent-infringement suit Gould and associates had brought against the company years earlier. Gould and his associates wound up with 54% of the company's stock, effectively taking control of the company.

Historians may never decide whether the Gould patents were truly appropriate.[4] But by the time LAA could begin addressing the issue, it was already too late to have any effect. The momentum against the industry was too great, and by August 1987, the LAA agreed that its only role would be educating the industry about the patents. At an industry-wide conference in January 1988,

[4]Bromberg, Lisa, *The Laser in America*, MIT Press, 1991, p. 75 ff.

LAA organized a seminar for companies where speakers discussed the inevitability of the Gould patents. During the ensuing months, virtually all the U.S. laser companies signed agreements honoring the Gould patents.

But the Gould patents were not the only issue facing the U.S. laser industry in 1987. Export controls, imposed by the government in the name of national security, were burdening the industry with tens of thousands of dollars in administrative costs, and were making U.S. lasermakers less competitive in the international market. Export controls had been a seminal issue in the formation of LIC, and were high on the priority list of the LAA in 1987. Chapter 5 is devoted to the industry's largely successful efforts over two decades to reform U.S. export controls on lasers and optics.

"Conference proliferation" was the catchphrase for the second major concern of laser and electro-optics manufacturers in 1987. As the number of universities and laboratories doing laser research grew, and as applications for lasers expanded, each of the laser-related professional societies expanded its conference schedule. Manufacturers felt compelled to participate in all the exhibitions held in conjunction with these conferences, lest an absence would be seen by potential customers as an indication of diminished competitiveness.

A single exhibition can cost a company tens of thousands of dollars in terms of personnel costs and shipping fees. From the manufacturers' perspective, it was far better to have a few large conferences/exhibitions in a year, than to have many small ones. But the trend was exactly the opposite: Each of the four professional societies was launching new, initially small conferences addressing different topics. Participation in all these exhibitions became a major expense in companies' annual budgets. Chapter 2 relates the tale of industry's dubious attempt during the next several years to alleviate this problem by consolidating many small conferences into one big conference.

But before either of these issues could be tackled, LAA membership had to be increased, and I was directed by the LAA board in June 1987 to make recruiting my first priority. Brochures were designed and printed, and board members were tasked to visit or telephone CEOs of nonmember companies to twist arms. And the promise of addressing two of the most pressing issues the industry faced was a potent recruiting argument. By the end of 1987, more than two dozen companies had joined the 16 LAA founders (Table 1.2).

By January 1988, I was able to increase my involvement to three-quarters time, and devoted my efforts to the three major LIC projects that year: recruiting, reform of export controls, and reduction of "conference proliferation." Members of the board were also heavily involved. John Wheeler, of Melles Griot, was named recruiting chair, and at the April 1988 LAA board meeting, set the goal of recruiting 25 new members, representing at least $25,000 in new revenue, during the year. By June, we had nearly a dozen new members, but after that the recruiting effort began to saturate. Five additional

**TABLE 1.2    More Than Two Dozen New Companies Had Joined LAA by the End of 1987**

| | | |
|---|---|---|
| Allied | Cascade Optical | Codman & Shurtleff |
| Continental Laser | Cryogenic Rare Gas | CVD Inc. |
| Diaguide | Directed Energy | ESI Inc. |
| Ferranti Electric | Image Engineering | KEI Laser |
| Kontes Glass | Koppers Co. | Labsphere |
| Laser Alignment | Laser Corp of America | Laser Machining |
| Laser Photonics | Laser Science | Lasermetrics |
| Liconix | Lumonics | Melles Griot |
| MIRA Inc. | Omnichrome | Quantrad |
| S.E. Huffman | Synrad | Questek |
| Two-Six | Wilson Ventures | XMR Inc. |

companies had signed up by the end of the year. LAA membership now comprised more than 60 companies, and several LAA board members calculated that LAA members manufactured at least 93% of the lasers manufactured in the United States.

Other LAA board members were taking a longer-term view, speculating on other projects the LAA might address after export controls and conference proliferation had been settled. A long-range planning committee was created, with Dean Hodges (see Figure 1.4) of Newport as its chair. The committee initiated a poll of members, asking about the industry's most-pressing needs. The potential project receiving the most positive response was a compilation

**FIGURE 1.4**    Dean Hodges was instrumental in LEOMA's creation and in guiding it through many undertakings.

of market data, so that companies could have a better perspective of the marketplace they served.[5]

Taking over from Jon Tompkins as LAA's president in 1989, Hodges convinced the board that the LAA should not limit its focus to lasermakers, but should also include a broad swath of companies involved in lasers and electro-optics. A new name was needed to emphasize the LAA's broader purpose.[6] After considerable discussion and a vote of the membership, the LAA renamed itself as the Laser and Electro-Optics Manufacturers' Association—LEOMA—at the June 1989 board meeting. Shortly thereafter, the charter was expanded to include "North American" laser and electro-optics companies, rather than U.S. companies.

As described in other chapters, the work with conference proliferation and export controls was moving rapidly. But all the time and travel associated with these projects was expensive, and the LEOMA found itself running out of money. For the fiscal year that ended in March 1989, we had spent nearly $80,000, but revenue from dues had been only $60,000.[7] LEOMA's reserves were shrinking at an alarming rate. Another issue, articulated by Mark Dowley of Liconix with the support of many smaller members, was the "nonlinearity" of the dues structure. While the absolute value of dues paid by larger companies was larger, smaller companies' dues represented a significantly larger percentage of their sales.

Treasurer Bob Gelber of Coherent proposed a major revision of the LEOMA's dues structure, which previously topped out at $5000 annual dues for companies with sales in excess of $20 million. But several companies—including Gelber's—had sales significantly in excess of $20 million. Under Gelber's plan, companies at the large end of the revenue spectrum, those with revenues in excess of $200 million, would see their dues increase 160%, to $13,000. At the lower end, the dues increase would be far less, only 10% for companies whose revenues were less than $6 million. From the entire membership, there was only one vote against Gelber's proposal, which went into effect in the summer of 1989. Although the structure still was not linear—dues for smaller members still represented a larger percentage of their sales—it was closer to linear than it had been. Dale Crane, the founder of Uniphase, was LEOMA's president-elect that year. In a recent interview, he reflected on

---

[5]Several board members argued that the calculation about 93% of the lasers being made by LAA members was dubious, because there were no hard data about the marketplace.

[6]"LEOMA" was not among the initial options. In the fall of 1988, the two leading candidates were "American Photonics Association" and "Photonics Manufacturers' Association."

[7]Each autumn, the board member designated as treasurer would work with me to design a budget for the following year. But day-to-day financial tasks—check writing, tracking budget categories, and so forth—fell to me. I made a detailed financial report at each LEOMA board meeting. I wrote and signed my own paycheck.

the size disparity between the two largest companies and the rest of the industry. "If [the dues] had been truly linear, Spectra-Physics and Coherent would have been paying for everything and the rest of us would have been coasting along for free."

And even as LEOMA was making headway with conference proliferation and export controls, another potential problem for the industry presented itself: "Europe 1992." The European Union was being formed, and along with a host of economic reforms, the Europeans were creating continent-wide standards organizations that would create standards for everything from automotive safety to screw sizes and included in the mix were new standards on lasers, laser optics, and other laser accessories. The industry viewed these new standards with alarm, and the LEOMA board was quick to add this issue to their association's agenda.

Initially, in the LEOMA board's view, the most efficient course would be to retain an attorney in Brussels, the seat of the European Union, to represent the U.S. laser and electro-optics industry in all matters European. That was an expense not anticipated in LEOMA's budget, but the LEOMA board viewed it as crucial. To cover the additional cost, the board passed a voluntary "standards assessment," effectively doubling the dues of those companies that agreed to participate. All the larger members did participate, and I was sent to Brussels, where I interviewed several attorneys who were eager to add LEOMA to their list of frightened U.S. clients.

But even as the LEOMA board was considering their various proposals, we were becoming more involved with the international standards bodies, the International Organization for Standards (ISO),[8] and the International Electro-technical Committee (IEC).

Chapter 3 describes LEOMA's successful efforts during the ensuing decade—and beyond—to influence the evolution of international laser standards.

The LEOMA board still identified recruiting as one of the association's most important activities. At the urging of LEOMA's 1990 president, Dale Crane of Uniphase, the LEOMA board members agreed in May 1990 to launch a major effort to recruit larger companies that use lasers, companies like HP, IBM, and others. These companies, the reasoning went, would be concerned with laser standards and export controls because they used so many lasers. A committee of past LEOMA presidents was tasked to design a plan for approaching these companies.

---

[8]It's incorrect to take ISO as an acronym, and call the organization the "International Standards Organization." Instead, "ISO" is Greek for "same" or "equivalent." The goal of standardization is to make measurements, procedures, and so on the same everywhere they're performed.

But at the next LEOMA board meeting, the past presidents reported that they were unable to design a viable plan for reaching these large companies. True, such companies may have been huge laser users, and they may even have been concerned with laser standards and export controls. But from their perspective, they wielded more political power by themselves than all of LEOMA put together could muster. They saw no benefit in joining LEOMA. "I can imagine the futility of that [recruiting] effort . . . now," Crane mused in a recent interview. But at the time, he and everybody else associated with LEOMA were intent on evaluating every growth mode possible.

At about the same time, LEOMA experienced another disappointing recruiting effort with a group of companies that manufactured laser machine tools. These tools are large instruments that use lasers to cut, weld, and otherwise process metals and other materials in automotive manufacturing and other heavy industries. The manufacturers of these tools wanted to have a trade association, and contacted LEOMA seeking information on how LEOMA might meet their needs. The board agreed that LEOMA could form a special section for these companies, and dispatched me to Chicago to deliver LEOMA's pitch at a conference of machine-tool builders.

But LEOMA's projects—standards, export controls, and conference proliferation—were not aligned with these companies' needs. They were interested in knowing how their products could penetrate an existing market that for decades had used conventional, non-laser, techniques for heavy manufacturing. Despite my assertion in the *Lasers & Applications* editorial nearly a decade earlier, this was not something with which LEOMA could help. I returned from Chicago empty handed.

Despite these recruiting disappointments, LEOMA was making substantial progress in its other projects. There was light at the end of the tunnels—or at least the end of the tunnel was in view, in the case of conference proliferation. Flush with these successes, the board began considering what challenges LEOMA could take on next.

The long-range planning committee put forth several ideas, including a market survey and enhancing the industry's interface with the federal government. Enhanced worker training was also discussed. But all these lacked the immediate urgency of the issues that had precipitated LEOMA's creation in the first place.

In September 1991, several members of the board and I visited Washington in search of inspiration for new LEOMA projects. We had appointments at the American Electronics Association (now AeA), the nation's largest high-tech trade association, where we hoped to learn about its activities that we might join or emulate. We also had appointments with several government agencies and departments, where we hoped to learn how LEOMA members could benefit by LEOMA's serving as an industry interface with the federal government.

At AeA, Bob Gelber of Coherent, who was LEOMA's president-elect that year, and I met with AeA president Dick Iverson and several other AeA officials. Iverson bent over backward trying to be helpful, and when our 11 AM appointment ended, he took us to lunch to allow an extra hour of conversation. He identified export controls and international standardization—two areas where LEOMA had already made significant headway—as issues of vital importance to any high-tech trade association. He explained that the AeA's interaction with the federal government was also important. But he identified the collection and distribution of market data as the most-appreciated function his organization performed. That, also, was a project that we had discussed at LEOMA, but after visiting AeA we realized that it could be a significant benefit to our members.

We visited several officials at the International Trade Administration (ITA) to evaluate how LEOMA might enhance its members' international sales through closer ties with the ITA. But we concluded that, while the international market was important to LEOMA, the ITA dealt with issues that were larger than the relatively small volume of sales in lasers and electro-optics. We saw no benefit to our members from interacting with the ITA.

One undertaking under discussion at LEOMA in 1991 was the possibility of organizing a research consortium among U.S. electro-optics companies. At the Commerce Department that September, LEOMA's 1991 president, Bob Pressley of XMR, and I met with several officials to discuss Commerce's support of such a project. The officials were very positive about industry's creating such a consortium, but they were not encouraging about Commerce Department funding.

The National Institute of Standards and Technology (NIST) is a part of the Commerce Department. While we were at Commerce, Pressley and I asked about NIST funding for LEOMA's work in international standard-ization, emphasizing that the United States was the only delegation at the ISO Laser Committee that lacked funding from its government. The best we could get from this discussion was a promise to look into the issue and get back to us.

When those of us who had visited Washington presented our findings to the whole LEOMA board, the reaction was mixed. Some board members were enthusiastic about launching new projects, while others were more dubious. But clearly, these proposed new projects lacked the urgency of the original issues LEOMA had been formed to address. In a memorandum to the LEOMA Executive Committee in late 1991, I summarized the question that hung over these deliberations: "Assuming that the issues of export control, international standards, and ["conference proliferation"] have been dealt with, does the laser/E-O industry still need a trade association?"

During the decade or so following its invention, the laser was often referred to, half jokingly, as "a solution in search of a problem." Now that the initial issues had been addressed, had LEOMA itself become a solution in search of a problem?

During the ensuing months, the momentum generated by LEOMA's successes in its original projects convinced its members that the industry did, indeed, still need a trade association. In the spring of 1992, the LEOMA board approved a new mission statement that identified several new directions for the association. Building from the results of our initial visit to the nation's capital, and especially our visit with the American Electronics Association, we would seek to establish constructive contacts with the federal government. Over the subsequent years, this project would yield several important successes, as described in Chapter 6.

The dubiousness of the previous autumn had been dispelled, and in an enthusiastic, unanimous vote, the board launched a project to create a quantitative study of the laser/electro-optics marketplace. The initial survey was distributed in May 1992. It was an overview of sales data that members had submitted, in full confidence, to the accounting firm Deloitte & Touche. Deloitte compiled the raw data and prepared a summary that described the overall marketplace, without including any company-specific information. Two years later, a second survey would be added to LEOMA's agenda, this one studying the compensation levels of engineers and technicians in the laser and electro-optics industries. The full story of these surveys, and of other intra-industry projects LEOMA addressed, is told in Chapter 7.

Meanwhile, international standards remained a concern, but one requiring much less effort than in previous years. Accordingly, the "standards assessment," begun in 1989, was discontinued in the spring of 1992. However, LEOMA's other projects were a drain on the association's assets, so a dues increase was passed at the same time. The larger companies' dues went up 15%, while the hike was smaller for smaller companies, all the way down to 5% for companies with revenue less than $600,000. That boost soon proved to be inadequate to fund all LEOMA's activities, and a second increase—this time 67% for larger companies, down to 15% for smaller companies, was approved before the end of 1992.

Of course, it was preferable to increase revenue by adding new, dues-paying members, rather than by increasing the dues for existing members. Newport's Randy Heyler, still leading LEOMA's recruiting effort, oversaw the creating of new recruiting materials emphasizing the new projects. He also solicited the two leading trade publications, *Photonics Spectra* and *Laser Focus World*, to run free advertising describing the association's new projects and their value to the industry. LEOMA was rewarded with five new members by the end of 1993.

Despite the dues increases in 1992, LEOMA's financial resources continued to diminish during 1993 as expenses associated with the new projects exceeded dues income. In early 1994, I proposed to the board that my short course, *Understanding Laser Technology* (ULT), become a LEOMA asset. ULT was (and is) a three-day course that I had developed long before LEOMA, but had ceased teaching recently because LEOMA took all my time. Now my concept was to begin teaching again, but the income from the course would go to LEOMA. The board accepted the proposal, and during the next six or seven years, I would teach the course a more than dozen times, both at LEOMA member companies and at public presentations around the country.

Buoyed by the extra income from ULT, LEOMA continued on a relatively even financial keel for the next several years. But there were emerging signs of trouble. Recruiting new members had slowed almost to a standstill. Several years earlier, David Rossi of Newport had distributed camera-ready copies of a small LEOMA logo to the membership, asking that members display the logo in a corner of their print ads. But in late 1995, a survey of the relevant magazines showed that only 4 out of 26 members' print ads included the logo. The board decided to discontinue the campaign.

To make matters worse, several small companies were dropping out. Most of LEOMA's projects—international standards, the interaction with the federal government—benefited the entire industry, not just member companies. In other words, companies could enjoy many benefits of membership without actually joining LEOMA and paying membership dues.

One incident in about 1997 was particularly telling. I was visiting a small laser company in Mountain View, trying to convince its CEO not to drop out of LEOMA. I argued that without LEOMA's participation in international standards, the Europeans would be free to create standards that could effectively block U.S. companies from the European market. "Spectra-Physics and Coherent are taking care of standards," he told me dismissively. What he meant, of course, was that Spectra-Physics and Coherent and other LEOMA members were paying LEOMA dues to protect all U.S. laser companies. My visit ended when he explained that his LEOMA membership cost as much as a new company-name sign on the front of the building. And he was going to opt for the sign.

Meanwhile, LEOMA's board was actively seeking new projects that would make the association more attractive, both to existing members whose loyalty was wavering and to potential new members. John Ambroseo of Coherent suggested launching a study of potential laser markets in the emerging field of extreme-ultraviolet lithography. Newport's Bob Phillippy suggested a campaign to reduce the burdensome requirement to obtain the European CE mark.

Neither of these ideas found much resonance with the membership as a whole. Other ideas were floated in late 1997 and early 1998. The National Fire

Protection Association, whose rules are often adopted by local governments, invited me to join its advisory committee addressing fires ignited by, among other things, lasers. I suggested that to the board, but it turned out that none of the members had experienced difficulty with fire regulations.

Recent federal legislation encouraged the creation of "Risk Protection Groups" of companies that could bind together and seek lower rates for liability insurance. Also, it was suggested that LEOMA make bulk purchase of magazine advertising space at a discount, and resell the space to its members. Neither of these ideas found favor with the board or with the individual members.

In the first months of 1998, two Canadian companies, Gentec and Lumonics, notified me that they would discontinue their membership. Both companies had been represented on the LEOMA board, and the absence of their dues would seriously undermine LEOMA's finances.

Still searching for appealing projects, I surveyed a dozen companies in the San Francisco Bay Area, and found that a remarkable shortage of laser technicians was likely to occur in the coming years. In 1998, these companies employed about 200 laser technicians, but by 2003, they predicted they would need at least 400. Where were these technicians to come from?

Chapter 4 describes the successful laser- and optics-technician programs LEOMA instituted at California community colleges. And while these programs were appreciated by regional companies, they did little to add to LEOMA's appeal for companies elsewhere. Indeed, their existence seemed not to encourage LEOMA membership even among regional companies. That same Mountain View company—the one with the new sign out front— actively competed with LEOMA members to hire graduates from the laser-technology program LEOMA had designed at San José City College.

But the absence of companies that had resigned from LEOMA in 1997 and 1998 was putting a severe crimp in the association's budget. Even with over $10,000 coming in from the ULT short course, my projection for 1999 was a shortfall of $40,000. Left with no alternative, treasurer Len Marabella and other board members worked out a budget that drastically curtailed many crucial LEOMA activities, but left it with a balanced budget for 1999.

That was not an acceptable solution to Spectra-Physics president Pat Edsell (see Figure 1.5). A long-time LEOMA board member and former LEOMA president, Edsell felt that curtailing these crucial activities undermined the association's fundamental reason for existence. Rather than allow the cutbacks to take effect, he offered funding from Spectra-Physics to underwrite LEOMA's entire $40,000 deficit.

Reflecting back on that 1998 decision recently, Edsell told me it was a worthwhile expenditure, even though in the end it merely postponed the inevitable. "I believe that LEOMA did the things a trade association should do. We can do more things better collectively than we can independently."

**FIGURE 1.5**    Pat Edsell, Spectra-Physics' CEO, was a long-serving LEOMA board member and the LEOMA president in 1994.

The industry needed LEOMA, Edsell said, and he was willing to do whatever he could to support it.

So LEOMA was out of the woods for another year, but it was clear that we still had to confront the question I'd posed seven years earlier: Absent the urgency of LEOMA's original, seminal projects, had LEOMA become a solution in search of a problem? To answer that question once and for all, LEOMA organized an industry-wide forum during a technical conference in San José in January 1999.

The forum was called "Solving Problems; An Alliance of Competitors," and was intended to bring non-LEOMA members into a discussion of industry-wide issues and their potential solutions. Many of these companies, as Spectra-Physics' Pat Edsell had pointed out in a *Laser Focus* editorial in December 1998, had withdrawn from such discussions because they considered themselves not "laser companies," but "semiconductor-equipment" companies or "telecom" companies or "medical" companies. But all these companies manufactured lasers and electro-optics, and therefore shared common problems that could be effectively addressed through LEOMA, Edsell argued.

The forum took its name from Andrew Procassini's book, *Competitors in Alliance*, cited on the first page of this chapter. And Procassini, who had led the Semiconductor Industry Association (SIA) for a decade, was one of two keynote speakers at the forum. Procassini credited the SIA with solving the major problems encountered by the U.S. semiconductor industry, and concluded that much of the credit for the U.S. leadership role in semiconductor technology went to industry executives who agreed to work cooperatively through the SIA to address common problems.

The second keynote speaker was Jon Tompkins, who while at Spectra-Physics had played a key role in LEOMA's early history. Tompkins had left Spectra and in 1999 was chairman of the Board of Trustees at KLA-Tencor and also chairman of SEMI/Sematech, the then-12-year-old industry association made up of majority U.S.-owned and -controlled chip suppliers. Tompkins described the importance of SEMI/Sematech to its industry and strongly urged the leadership of the U.S. laser and electro-optics industry—most of whom were in that room—to support LEOMA.

Following the keynote talks, a panel of six industry leaders addressed a pair of crucial questions: What were the most urgent issues facing the industry as a whole, and how can companies most effectively address those issues? The six panelists had been chosen for their experience in the laser/electro-optics industry, and for the diversity of their opinions. Of the six, only two—Bernard Couillaud, president and CEO of Coherent, and George Balogh, VP and general manager of Spectra-Physics' optics division—were affiliated with LEOMA member companies. The other panelists were David Rossi, VP of marketing in Opto-Sigma; Lindsay Austin, VP and general manager of Uniphase's laser division; Don Scifres, president and CEO of SDL; and Bob Mortensen, president and CEO of Lightwave Electronics. Dave Hardwick, LEOMA's 1998 president, moderated the discussion.

The LEOMA members of the panel argued that the industry's crucial issues—international standards, government regulations, worker training, and so forth—were precisely those issues that LEOMA was addressing. The counterargument was voiced forcefully by Lightwave Electronics' Mortensen, who held that when a truly urgent issue confronted the industry, companies could unite to address it. But in the absence of potentially catastrophic developments, he insisted, a trade association was an unnecessary expense.

But Mortensen was in a definite minority, and the LEOMA board convened as a newly invigorated body at its next meeting. The last year of the century would be the year to replant LEOMA and revitalize the association as an integral part of the photonics industry. An ambitious plan to restructure LEOMA was launched, with committees to address each of the prime objectives the board had identified. Each committee consisted of three to five board members, one of whom was designated as the chair.[9] The concept was that each of these committees would focus on its particular project, being more efficient than the entire board, whose attention in past years had been diluted over LEOMA's entire scope.

During 1999, these committees made a serious effort to cope with their respective projects, but the truth was that each committee member was a high-level manager at his own company, and company concerns outranked LEOMA concerns. By midyear, the concept of action committees composed of board members was wavering, and by the end of the year it had effectively been abandoned.

And while the LEOMA supporters—board members and others—were recharged by the "Alliance of Competitors" meeting in January, the enthusiasm had not spread well into the larger community. Few new members were recruited, and LEOMA was nowhere near generating additional dues income to replace the $40,000 that Pat Edsell had supplied to underwrite LEOMA's 1999 activities. In putting together a budget for CY 2000, I increased the contribution from ULT to $20,000, and still predicted a shortfall of $30,000. LEOMA had sufficient resources to absorb the shortfall, but it clearly was not a steady-state situation.

And further problems began surfacing. It turned out that not only was the concept of action committees composed of board members impractical, but also board members were so involved with issues at their individual companies that none of them was able to serve as LEOMA secretary during 2000. Minutes during 2000 were taken by an individual (most often a board observer from *Photonics Spectra* magazine, rather than a regular board member) drafted into service at the beginning of each meeting.

Another setback in 2000 was the demise of the LEOMA Marketplace Survey. As explained in Chapter 7, despite the enthusiasm with which the board had initiated the survey a decade earlier, the project had not been terribly successful during the ensuing years. At the October 2000 board meeting, I proposed discontinuing the survey, and the board agreed to do so.

Yet another blow during 2000 was the resignation from LEOMA of one if its prime members, SDL Inc. After absorbing a $30,000 shortfall in 2000, and now accounting for the absence of SDL's dues in 2001, LEOMA was looking at a shortfall of $50,000 for that year. LEOMA's diminished resources were inadequate to cope with a deficit that large.

---

[9]The committees were (1) Government affairs, chaired by Spectra-Physics' Pat Edsell; (2) Bylaws revision, chaired by Tom Cekoric of Applied Optronics; (3) Manpower and training, chaired by Newport's Bob Phillippy; (4) Small-company projects, chaired by Chong Lee of Lee Laser; (5) Export control, chaired by Dave Hardwick of Galileo Corporation; and (6) Surveys, chaired by Len Marabella of TRW.

But these setbacks were at least partially balanced by the progress LEOMA was making with worker-training programs. The predicted shortage of engineers and scientists—in all fields, not just photonics—was perceived as a strategic national problem. The shortage was especially acute in the optics field,[10] where the unprecedented growth of fiber-optics technology created a seemingly insatiable demand for optical components. Chapter 4 describes how LEOMA had begun addressing the problem years earlier. Those efforts were now bearing fruit with successful programs at several community colleges, with the PowerPoint presentation prepared by Bob Phillippy describing careers in optics, and with my own stint as chair of the Coalition for Photonics and Optics,[11] where I was moving that institution to focus on programs encouraging high school and college students to consider careers in optics and photonics.

Nonetheless, the $50,000 cash shortfall predicted for 2001 was an overwhelming cloud on the horizon. I feared that LEOMA would be forced to discontinue operations despite its successes in worker training. But in a turn of events I had not expected, the board eliminated that cloud in less than 10 minutes during its October 2000 meeting.

Dave Dover of *Photonics Spectra*—who had "volunteered" to take the minutes of that meeting—said that Wendy Laurin, the magazine's publisher, had instructed him to announce that the magazine would contribute $5000 to LEOMA's treasury as a step toward alleviating LEOMA's financial plight. Steve Sheng then said Spectra-Physics would donate $10,000 in addition to its normal dues, and John Ambroseo made a similar donation from Coherent. Bob Phillippy of Newport added another $10,000, as did Mike Dorich of Melles Griot. This amounted to $45,000 in donated funds against the $50,000 shortfall, and Steve Sheng then said Spectra-Physics would come up with the remaining $5000 if nobody else did.

So, for the second time in three years, LEOMA's significant operating deficit would be underwritten by a few of its largest members. At its January 2001 meeting, the board began anew the effort to find the formula for a successful trade association in the laser/photonics industry. Recruiting new companies was an obvious priority, and board members themselves committed to contacting nonmember companies in an effort to learn what might attract them to LEOMA.

Scott Keeney of nLight, who was an active board member, contacted several companies including Aculight and Phaethon, finding "lukewarm"

---

[10]The HR manager at one large optics company told me she would visit fast-food restaurants and, if anybody behind the counter looked particularly intelligent, make a job offer on the spot.
[11]CPO was a coalition of professional societies and other organizations described in Chapter 2.

interest at best. These companies' executives were "incredibly busy," Keeney reported, and not particularly interested in industry-wide issues.

Steve Sheng of Spectra-Physics, LEOMA's president in 2001, spoke with officials at New Focus and Lightwave Electronics, and found those individuals too preoccupied with internal issues to be interested in discussing LEOMA.

John Ambroseo of Coherent, another active board member, contacted several additional companies. At Corvis, he found general disinterest in any of LEOMA's projects. At Avanex, he found his contacts too focused on many other short-term issues to focus on the benefits of a trade association.

At a technical conference earlier in 2001, Jeff Canon of JDSU had told me that he was enthusiastic about participating in LEOMA, but that enthusiasm evaporated a month later when I formally approached him. I also spoke with SDL's CEO Don Scifres, whose resignation from LEOMA the previous year had precipitated a financial crisis. But he said that "budget pressure" would prevent his rejoining the association in the foreseeable future.

So, all in all, the prospects for meeting LEOMA's financial shortfall by recruiting new members were not favorable.

A proposal was made at the May 2001 board meeting to readjust LEOMA's dues schedule to reflect the greater benefits to California companies. All LEOMA's community college programs were in California. But California companies were already providing more than three-quarters of LEOMA's income, so the board nixed any revision of the dues schedule.

During that May meeting, board members tabled several ideas for lifting LEOMA out of its financial straits. It was suggested that we contact other associations, like the Semiconductor Industry Association and the National Machine Tool Builders' Association, to learn if any of their "best practices" might be something LEOMA could emulate. Once again, the possibility of increasing membership surfaced. Prefacing his comment by saying he didn't want to sound like a curmudgeon, Pat Edsell observed that all these approaches had been tried before, unsuccessfully.

Edsell was right. Reluctantly, the board began examining LEOMA's projects with an eye to abandoning the least crucial ones. The executive seminars, annual meetings of the board and other executives of LEOMA companies, had been held in Washington in recent years. They were viewed as a valuable opportunity to network with executives from other companies, and an important part of the interface with the federal government. But they cost time and money to organize, and they appeared less vital than LEOMA's other activities. The board voted to discontinue them.

Time and money were also involved in organizing another set of seminars, the Human Resources seminars for LEOMA's HR managers. Although the HR managers found these seminars useful, they did not provide a vital contribution to the industry as a whole. The board instructed me to inform the HR directors at various LEOMA companies that LEOMA would no longer organize the seminars. If they wanted to continue them, they would have to organize them themselves. Of course, the board knew that this was unlikely and, indeed, there were no further HR seminars after 2000.

Finally, the board considered the possibility of a merger with one of the professional societies. Feelers went out to two of them, the Optical Society of America (OSA), and the International Society for Optics and Photonics, known by an acronym for its former name, SPIE. The OSA responded positively toward the end of 2001. OSA had a strong program of short courses presented at conferences, and they were interested in LEOMA's community-college programs. Their strongest interest, though, was in LEOMA's 501(c)(6) tax status. As a 501(c)(3) corporation, OSA was barred from many political activities that LEOMA could perform. The OSA concept was that LEOMA would retain its (c)(6) status and become a subsidiary of OSA.

There were advantages to LEOMA of such a merger. OSA had a Corporate Associates program, which, while far less active than LEOMA, had many more members. The idea was, if the merger occurred, LEOMA could absorb OSA's Corporate Associates, thereby boosting its income and, in the process, making its services available to a wider swath of industry.

A conference call in November involved the presidents and executive directors of LEOMA and OSA, as well as Duncan Moore, who was the OSA's senior science advisor, and LEOMA board members Bob Phillippy and John Ambroseo. Dave Hardwick, who was then an OSA board member but was also a past president of LEOMA, also took part in the conversation. All participants agreed that a merger would in many ways be advantageous for both institutions. But OSA wanted LEOMA to operate out of its Washington, DC, offices, and I did not wish to relocate to the East Coast.

That turned out to be a deal breaker. With two kids in grade school and a wife successfully running her own business in California, I was unwilling to uproot the whole family and move it to Washington. And the LEOMA board was equally unwilling to turn day-to-day operation of LEOMA over to a stranger from the OSA. Shortly after the conference call, LEOMA formally asked OSA to put further discussions "on hold."

But another possibility soon emerged when SPIE responded positively to the feelers LEOMA had generated. As had been the case with OSA, SPIE was interested in LEOMA's 501(c)(6) tax status. In December 2001, I met Eugene Arthurs (see Figure 1.6), the SPIE executive director, at the San Francisco airport and we drove together to Mountain View, where we met with LEOMA's president, Steve Sheng, and president-elect, John Ambroseo.

Arthurs explained that SPIE was interested not only in LEOMA's tax status, but also in access to the LEOMA companies, whose leaders had for

**FIGURE 1.6**  Eugene Arthurs, the SPIE executive director, was interested in absorbing LEOMA, but after its initial enthusiasm, the LEOMA board decided against joining forces with SPIE.

years been championing solutions to industry-wide issues. And he offered to help LEOMA expand its community-college program nationwide, using SPIE's network of student chapters as a basis. The mechanics would be similar to those envisioned earlier in the OSA discussions: LEOMA would retain its own board and its (c)(6) tax status, and become a subsidiary of SPIE. But unlike OSA, SPIE had no issue with LEOMA's keeping its headquarters where they were in California. By the end of the meeting, both parties agreed to take the proposal to their respective boards in January, and to meet for further discussions at an SPIE technical conference in late January. If both boards were in agreement, the wording of a contract could be achieved by summer, and the formal merger could take place by September.

Some skepticism to the merger plan surfaced at the January 2002 LEOMA board meeting. One question was whether LEOMA should retain its name after the merger. Several board members felt it would be beneficial to drop the name "LEOMA," because most companies knew—or thought they knew— what LEOMA was and had made up their minds about becoming associated with it. But others worried that losing the name would lead to LEOMA's being subsumed into SPIE, even though it retained its own board.

But what were the options? LEOMA had gotten through 2001 on the strength of donated funds from its leading members, but those companies were not willing to continue that level of support. At the board's request, I fashioned a proposed budget for 2002 that showed a balanced budget without curtailing any LEOMA activities. The donated funds from 2001 would be replaced by income from LEOMA's short courses. LEOMA had been receiving income from *Understanding Laser Technology* for several years, and meanwhile I had added a second course, *Understanding Fiber-Optics Technology* (UFT). My proposed budget called for a dozen presentations of the courses during 2002, six presentations of ULT and six of UFT, bringing in an additional $50,000.

Were that many presentations feasible? A quick survey of companies represented at that board meeting indicated that nine presentations could be hosted by those companies alone. Emboldened by the possibility of continuing independent operations, the LEOMA board instructed its representatives— Steve Sheng, John Ambroseo, and myself—to slow down the pace of those negotiations during the subsequent meeting with SPIE.

At the last minute, Sheng had a family emergency that prevented his participation at that meeting, and Pat Edsell asked if I wanted him to fill in. I gratefully accepted the offer. But at that pivotal meeting, Edsell went a lot further than "slowing the pace" of merger negotiations. He single-handedly torpedoed the merger. LEOMA didn't need SPIE's support, he insisted, and a merger would dilute LEOMA's strength. By the end of that meeting, a merger between SPIE and LEOMA was no longer a possibility.

Ironically, neither Edsell nor Ambroseo today have any recollection of that meeting. But one person who does remember it is Eugene Arthurs, SPIE's executive director. "When LEOMA approached us, we formed a subcommittee of the SPIE board to evaluate the matter," he told me recently. "We decided it was promising, and after the discussion you and I had, the subcommittee came to the meeting with LEOMA." They were taken aback by the resistance to the idea they encountered at that meeting. Arthurs found the dénouement of the negotiations "insulting," especially since LEOMA had approached SPIE in the first place.

Speculating recently on his motives at the meeting, Edsell mused that he didn't want LEOMA to be absorbed into SPIE and have its impact diluted. "SPIE does a lot of things," and LEOMA's projects would be low on the priority list. He "felt strongly that LEOMA was important, and that industry [not a professional society] should support it." But LEOMA did go out of business shortly after that meeting, I pointed out. "Maybe I was wrong. I've made a few mistakes in my career." But, he insisted, LEOMA operating as a subsidiary of SPIE would have been of very limited value to the industry.

But the optimism that had fueled the notion in January that LEOMA could fund operations through 2002 with income from the short courses proved unfounded. The fiber-optics boom that had driven rapid growth in the optics/photonics industry during the last years of the previous century came crashing down in the first years of the twenty-first century. By early 2002, demand for optical components had diminished drastically, and companies that had previously been desperate for employees were suddenly laying people off. LEOMA's optics program at Yuba College came to a screeching halt, and its recent graduates who had found new jobs at optics companies were often the first to be let go. The demand for LEOMA's new short course, *Understanding Fiber-Optics Technology*, shrank to zero, and by mid-2002 it became obvious that the short courses were not going to generate the income to keep LEOMA in the black for 2002.

At the May 2002 board meeting, the discussion focused on whether LEOMA's large members would—for the third time in four years—provide the emergency funding to keep the association going. But John Ambroseo, LEOMA's 2002 president, observed that one thing he'd learned in his years at Coherent was, "If it's not working, quit throwing money at it."

LEOMA was not working. The entire industry was in a slump, and LEOMA dues were among the lowest priorities on companies' lists. One of LEOMA's most successful undertakings, training programs for optical and laser technicians, was producing technicians who could not find employment. And I had begun reversing the process of 14 years ago, making commitments of my time to clients other than LEOMA. By the end of 2002, nearly half my time was devoted to non-LEOMA projects.

The board agreed to put LEOMA in a "simmer mode," where only the association's crucial activities would receive minimal maintenance funding. The activities were defined as export controls, international standards, the ADR[12] agreement, and the Coalition for Photonics and Optics. All LEOMA's education-related activities were halted. There would be no more recruiting, no more seminars or surveys, no more interfacing with the federal government.

But the "simmer mode" is not a long-term strategy. Interest in LEOMA's issues continued to wane until, at its January 2005 meeting, the LEOMA board agreed to discontinue operations as a dues-collecting industry association and transform into a loose confederation of companies, still calling itself "LEOMA," with a paid consultant. Ambroseo proposed, and the board unanimously agreed, to transfer the balance of LEOMA's treasury to me as a "severance package." From that point forward, the companies paid the consultant—me—directly, rather than paying LEOMA dues.

And that arrangement has survived to the present day. The LEOMA companies continued to hold "board" meetings from time to time, which

---

[12]The Alternative Dispute–Resolution agreement is described in Chapter 7.

in reality were merely meetings of an advisory committee. In 2006, Michael Lebby, then executive director of the Optoelectronics Industry Development Association (OIDA), visited one of these meetings in an unsuccessful attempt to recruit the LEOMA companies to OIDA. Chapter 5 describes the LEOMA-driven revision of international laser export controls from 2001 to 2006. In 2006, Jim Harrington of the State Department visited one of LEOMA's "board" meetings to describe the newly revised controls and praise LEOMA for its contributions to them.

## THE BENEFIT OF HINDSIGHT

In the end, LEOMA did indeed turn out to be a solution in search of a problem. LEOMA was created to address two very specific problems: export controls and conference proliferation. Shortly after its founding, the association identified a third issue, international standards. LEOMA was quite successful in dealing with export controls and international standards and, to be fair, at least partially successful in dealing with conference proliferation.

The momentum of those successes carried LEOMA forward for the next decade. During that decade, the association undertook a variety of new projects, from market surveys to worker training to interfacing with the federal government. Most of these projects achieved their intended goals, but the urgency of the initial three projects was never repeated. Much of the initial momentum was gone by the turn of the century, and the industry downturn that accompanied the new century proved to be LEOMA's undoing. When budgets were being cut, LEOMA's less-than-urgent projects were among the first to go.

Even before the momentum began to diminish, LEOMA suffered from the "public television" problem: Many benefits aren't dependent on membership. Just as all viewers can watch public TV programs, all laser companies benefited from most of LEOMA's activities, whether or not they joined LEOMA. Development of international standards, reform of export controls, training programs for industry workers—all of these projects benefited the entire industry, not just those companies that paid LEOMA dues. Many companies calculated, correctly, that they would derive more self-benefit from spending their money elsewhere. Projects whose benefits accrue exclusively to the members are crucial in a trade association's success.

With the few projects LEOMA undertook whose benefit could have been exclusive, the tendency was to open them to nonmembers. The executive seminars, the market survey, the compensation survey—these projects could have been for members only, but in most cases nonmembers were also invited to participate. The motivation behind opening these projects to nonmembers was increased industry participation. The more companies that participated in the surveys, the more reliable the resulting data would be. The more companies that attended the executive seminars, the more valuable the networking would be. And, to be frank, the more companies participating in—and underwriting—these projects, the better for LEOMA's bottom line.

The reality was that very few nonmember companies participated in any of these projects. In hindsight, one wonders if the perceived value of these projects might have been greater if they had remained exclusive. The executive seminars, in particular, had a certain degree of cachet. As Randy Heyler succinctly explained in a 2013 conversation, "The LEOMA executive seminars were a great thing, because it played to people's egos. People could say, 'I'm an industry leader, so I get to go to this seminar and talk to other industry leaders.'" Perhaps that cachet would have been greater if the seminars had been exclusive. Perhaps, if these projects had been members only, they would have encouraged more companies to investigate LEOMA membership.

# 2

# PROFESSIONAL SOCIETIES AND THE PHOTONICS COMMUNITY

In the United States, the "photonics community" is a loosely defined entity consisting of people and companies working in one way or another with lasers and photonics. It includes companies that make lasers and related optical equipment, and the engineers, technicians, scientists, and academics working in the field. It is a diverse community with different and often competing interests. In the past quarter century, there have been two major attempts to bring the various elements of the community together for the common good. The first attempt, in the late 1980s and early 1990s, was a conference called OPTCON; the second, in the late 1990s and the first years of the twenty-first century, was the Coalition for Photonics and Optics.

## OPTCON

The first of those, OPTCON, was intended to be the answer to "conference proliferation." A 1987 poll of U.S. laser manufacturers found that the industry's most pressing problem was the rapid growth in the number of technical conferences and the expense related to mounting exhibits at the conferences. OPTCON was to be a conference that brought together, at the same time in the same place, many of these proliferating conferences.

*LEOMA and the U.S. Laser Industry: The Good and Bad Moves for Trade Associations in Emerging High-Tech Industries*, First Edition. C. Breck Hitz.
© 2015 by The Institute of Electrical and Electronics Engineers, Inc. Published 2015 by John Wiley & Sons, Inc.

"It seemed like a good idea," Gary Bjorklund reminisced recently. "There were certainly a lot of things going on in different places. It seemed to make sense to bring them together." Bjorklund, a former IBM researcher, served for years on the OPTCON steering committee, and chaired the committee in 1990.

But OPTCON was to be more than a consolidation of existing conferences. Most of the laser and electro-optics conferences that existed in 1987 were attended by researchers and academics, discussing their latest laboratory results. OPTCON's focus, instead of research, was to be applications of lasers—in manufacturing, medicine, sales, communication, and elsewhere. OPTCON would bring together a "new audience" of engineers and technicians describing their use of lasers in all these fields.

The "proliferating" conferences were sponsored by the four optics-related professional societies, which in 1987 were (as they remain today) the Optical Society of America (OSA), the IEEE Laser and Electro-Optics Society (LEOS, now the IEEE Photonics Society), the Laser Institute of America (LIA), and the International Society for Optics and Photonics (SPIE). LAA's Glenn Sherman, president of Laser Power Optics, invited the executive directors of all the societies to meet with him in January 1987 to discuss consolidating many of their existing conferences into one large conference. Sherman's pledge of industry support for the new conference was a strong inducement. The budding LAA concept, he explained, was to hold a large laser and optics conference in the fall, separated by six months from the community's then-dominant conference, the Conference on Lasers and Electro-Optics (CLEO). The new "fall conference" would highlight applications, in contrast to CLEO's focus on research.

The concept made sense to most of the societies, but SPIE saw it as conflicting with its own interests. SPIE had always had a less-academic focus than the other societies, and its meetings tended to place more emphasis on applications. One of its meetings in particular, O-E Lase, sought papers discussing applications of lasers and electro-optics.[1] Joe Yaver, the executive director of SPIE at the time, recalled recently that he thought being asked to support the new applications-oriented conference was like "being asked to put a bullet in our own ambitions."

Despite his reservations, Yaver and SPIE became a sponsor of the new conference. "Everybody wanted cooperation among the societies," Yaver recalled. The SPIE board comprised individuals who had close connections to the other societies, and they convinced Yaver to accept the concept. By the

---

[1]O-E Lase continued seeking applications-oriented topics during the ensuing years. It changed its name to Photonics West and moved from Los Angeles to San Jose to San Francisco, and by the early 2000s had become the best-attended photonics conference/exhibition in North America.

middle of 1987, the four societies and LAA had agreed to jointly sponsor a fall conference whose emphasis would be on applications of lasers and electro-optics.

But getting agreement on the concept of a "fall conference" was one thing, organizing it was another. Who would be in charge? Who would book the exhibit hall? Who would sell exhibit space? Who would organize the short courses, employment center, press room, and all the other components of a modern technical conference?

As a first step, the five institutions formed a conference steering committee with two voting representatives from each. LAA's Dean Hodges, a Newport employee who had been instrumental in the initial discussions for a fall conference, chaired the committee. And in late 1987, the new steering committee abruptly faced a major decision: space was available to accommodate the conference in less than a year at the Santa Clara (California) Conference Center.

The dates were from October 30 to November 3, 1988. It was time to fish or cut bait. None of the details of the new conference had yet been worked out. Should the steering committee plunge into this undertaking and assume it could work out the details in the coming months? Taking a deep breath, the committee voted to do so.

One of the first details was to find a name the conference, which at the beginning of 1988 was still being referred to as the "fall conference." The steering committee unanimously agreed to "OPTCON," short for "optical conference."

And the Optical Society, with an experienced meetings department, took on responsibility for most of the organizational details of the first OPTCON, including selling exhibit space. Each society was to organize its own technical sessions. In most cases, those sessions would be the society's annual meeting, moved to take place at OPTCON. There would also be an "OPTCON Core Program" consisting of the exhibition, short courses, daily plenary lectures and panel sessions, and an employment center.

LAA assigned itself the task of ensuring that OPTCON has a strong applications orientation. "OPTCON is not CLEO" was to be the mantra. Jon Tompkins, the CEO of Spectra-Physics and the 1988 president of LAA, composed a letter sent to all potential exhibitors encouraging them to display non-research products. "The target audience is OEMs [original equipment manufacturers], not professors," the letter declared.

The steering committee identified three specific applications to be emphasized at OPTCON '88: semiconductor fabrication, biotech, and military. LAA assigned board members to organize a technical session for each topic as a part of the core program. Marty Cohen of Quantronix agreed to organize a panel with three speakers—a representative from a laser manufacturer, a representative from the semiconductor industry, and a journalist—to discuss the use of lasers in semiconductor manufacturing. Gary Goodman of Laser Power would organize a similar panel on military applications, and Vittorio Fossati of Coherent would do likewise with biotech.

The LAA also organized a series of "how-to seminars" aimed at laser manufacturers who would exhibit their wares at OPTCON. Topics included: how to design an advertising campaign, how to establish an overseas distribution network, how to manage customer service, and other topics. OPTCON's target audience was OEMs, so LAA created private meeting spaces at the edge of the exhibit floor where potential laser users could meet in confidence with laser engineers to discuss the feasibility of their ideas. Finally, LAA organized a tutorial "State of the Technology" session, where engineers from various LAA companies would discuss—without sales pitches—the characteristics of different types of commercial lasers.

By September 1988, all the hotel rooms reserved for conference attendees had sold out. All 88,000 square feet of exhibit space had been sold. Despite its short planning period, OPTCON '88 was looking like it would be a success!

And it was. By the time the counting was finished, OPTCON '88 had attracted over 7500 attendees (see Figure 2.1). Some 350 companies had exhibited lasers and electro-optical equipment, and the 50 short courses had been well attended. Most important, OPTCON '88 had been profitable for its sponsors, netting more than $300,000 revenue to be divided among them.

A survey of exhibitors showed that 60% of those companies believed that OPTCON was reducing conference proliferation (but 40% believed it was not).

FIGURE 2.1   The Coherent booth being set up at the first OPTCON, at the Santa Clara Convention Center, in 1988.

Fifty percent saw a "new audience" of laser users, rather than the researchers they saw at CLEO, and 70% described themselves as "pleased overall" with the conference. But despite these signs of success, most exhibitors ranked CLEO as "somewhat better" than OPTCON.

So the stage was set for continuing OPTCON as the new fall conference. Efforts had begun in early 1988 to secure a location for OPTCON '89, but it developed that only Orlando had suitable facilities available in the fall of 1989. And SPIE objected strongly to siting a fall conference in that city, because it would compete with an existing SPIE conference in Orlando. Reluctantly, the steering committee decided to forego an OPTCON in 1989 and move directly into planning OPTCON '90.

Space for OPTCON '90 was secured for the week of November 5 at Boston's Hynes Auditorium. The steering committee agreed that SPIE, which also had an experienced meetings department, would assume the overall managerial responsibilities that OSA had discharged in 1988. Several logistical problems immediately emerged: parking would be difficult around Hynes; the exhibition would have to be split between two floors; both labor and floor space were more expensive than they had been in Santa Clara. Floor space, for example, was nearly four times as expensive as it had been for OPTCON '88. On the other hand, proponents argued that OPTCON '90 would be the largest laser/electro-optics exhibition ever held in New England, and would open vast new markets.

LEOMA[2] set out to convince its members to mount major exhibits at OPTCON '90. Dean Hodges calculated that the ratio of attendees to exhibition floor space had been steadily decreasing at CLEO in recent years, and argued that LEOMA members should scale back at that conference and scale up at OPTCON.

To swell the attendance at the Boston meeting, LEOMA members provided highly proprietary mailing lists of their customers to direct-mail houses, which kept the lists carefully guarded but orchestrated mailings to thousands of customers throughout New England. LEOMA Board members also organized conference activities similar to those at OPTCON '88. A seminar describing the international laser marketplace would feature the CEOs of LEOMA's largest members as speakers. The how-to seminars, which had been very popular in Santa Clara, would be repeated in Boston.

But as its sponsors were preparing for OPTCON '90, they were wrestling with what they wanted the conference to evolve into in the long term. A poll of LEOMA members indicated that 84% supported continuing OPTCON, 16% were neutral, and nobody opposed. But the professional societies, which had

---

[2]The organization changed its name from Laser Association of America to Laser and Electro-Optics Manufacturers' Association in June 1989.

initially colocated their annual meetings with OPTCON, decided they wanted to discontinue that practice. Jarus Quinn, the executive director of the Optical Society, announced that the OSA Annual Meeting would separate from OPTCON in 1991, and LIA president Murray Penney followed suit. Both societies pledged to organize technical sessions at OPTCON, but their annual meeting would be separate. Bob Wangemann, the LEOS executive director, said LEOS would keep its annual meeting with OPTCON in 1991, but would probably withdraw in 1992.

OPTCON's long-range planning committee, which by that time was chaired by Dean Hodges, recommended that OPTCON evolve into a pure trade show with a large equipment exhibition focusing on applications, but no technical sessions. Gary Bjorklund, who chaired the steering committee in 1990, listed four possibilities:

1. OPTCON could stay the course, with heavy loading of technical content from the professional societies.
2. OPTCON could downsize, but maintain small technical sessions organized by the societies.
3. OPTCON could colocate with a larger technical conference like WESCON.
4. OPTCON could evolve into a stand-alone trade show with no technical sessions.

Although many on the steering committee were doubtful that such a stand-alone trade show could survive financially, in June 1990, the steering committee formally approved the recommendation of the long-range planning committee.

Even as the steering committee was deciding to change the very nature of OPTCON, anxiety about OPTCON '90 was increasing. While people could speculate that the conference would draw many attendees to Boston, the audience in the northeast corner of the country was untested. The LEOMA board quietly decided that, if OPTCON '90 were unsuccessful, LEOMA's sponsorship of the conference would end immediately.

In the last weeks of October, everything appeared to be going smoothly. Preregistration numbers were good, and at a steering committee meeting on the eve of the conference, SPIE was awarded a round of applause for its successful organization.

And OPTCON '90 was, by some measures, quite successful. The total attendance of 7,445 was nearly as great as that at OPTCON '88, most attendees and exhibitors were enthusiastic about the conference, and by

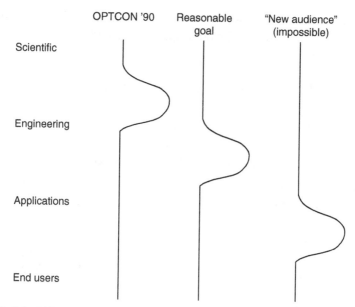

**FIGURE 2.2** Milton Chang argued that an audience of end users was unrealistic for OPTCON.

the time the accounting was finished, the net revenue of $856,000 was twice what it had been in Santa Clara.

But by other measures OPTCON '90 was disappointing. SPIE conducted an informal poll of exhibitors that indicated only 7% of the attendees could be considered the new applications-oriented audience OPTCON was designed to attract. And as the professional societies pulled their annual meetings out, OPTCON's goal of diminishing conference proliferation became increasingly dubious.

Bill Silfvast, the incoming chair of the steering committee for 1991, inherited a committee roiled by uncertainty. At the January meeting, several people questioned the reality of attracting a "new audience" composed of applications-oriented engineers. Milton Chang articulated this concern with three curves (Figure 2.2). As applications matured, Chang argued, the people developing those applications would go to conferences focusing on the applications, not conferences focusing on lasers.

SPIE still viewed OPTCON as an intrusion into its territory of applications-focused conferences, and in the confusion now surrounding the future of OPTCON they saw an opportunity to end it. At the January 1991 meeting, SPIE representative Bob Sprague formally moved that OPTCON be discontinued after the 1991 meeting. Taken aback by the abruptness, the other four sponsors voted against the motion.

Sprague then declared that SPIE would withdraw as an OPTCON sponsor after the 1992 meeting unless the other sponsors agreed to a radical restructuring of the conference:

- SPIE would design the technical program (taking into consideration suggestions from the other sponsors).
- Income from the exhibition, attendee registration, and short courses would be evenly divided among the five sponsors.
- All sales commissions from exhibition would go to SPIE.
- Eighty percent of the income from publications would go to SPIE.
- OPTCON would stay in Boston every year, managed by SPIE.

Voicing the apparent reaction of the other sponsors, Chang expressed "shock" at the audacity of the SPIE proposition. Sprague acknowledged that it was a "radical" shift, but insisted it was the only way SPIE could continue as an OPTCON sponsor.

Chairman Silfvast called a recess and asked the SPIE contingent to leave the room. During the recess, the other four sponsors quickly agreed that they could not accept the SPIE proposition without consulting their respective boards. Silfvast called SPIE back into the room, and directed the executive directors of all five sponsors to negotiate a plan for OPTCON to go forward after 1991 with or—if necessary—without SPIE.

From the LEOMA perspective, OPTCON was evolving into something far different than had been envisioned three years earlier. Instead of reducing conference proliferation, it was *adding* a conference to the list of exhibitions where companies felt compelled to display their wares. Moreover, the poll indicating that only 7% of the audience was "new" was discouraging. And the discord among OPTCON's sponsors lent a very sour note to the whole project. At the March 1991 LEOMA board meeting, President Bob Pressley attempted to distill straightforward options from the confusion, explaining that there were three options: (1) LEOMA could accept the SPIE proposition, (2) it could undertake to continue OPTCON without SPIE, or (3) it could drop out of OPTCON after 1991. With little discussion, the board selected the third option.

It was a decision that lasted barely two months. Careful surveys of exhibitors subsequent to OPTCON '90 painted a far rosier picture of the situation than the perhaps overhasty initial survey SPIE had conducted. According to these new surveys, 91% of the exhibitors felt OPTCON "was important to E-O industry." And 77% felt OPTCON was bringing its intended "new audience." These numbers were significantly different from the informal survey taken on the floor of the conference. Although there was uncertainty about which numbers were closer to reality, the LEOMA board at its May 1991 meeting voted to reverse its decision to end OPTCON sponsorship.

As instructed by the OPTCON steering committee, the executive directors of the five sponsors attempted to find a path for OPTCON to move forward after 1991. OPTCON '92 would be held in Boston, and SPIE had previously agreed to be the conference manager. It was agreed that SPIE would sign a contract for Boston's Hynes Auditorium for November 1992, and all five sponsors would share any loss that resulted from a cancellation of OPTCON '92.

Meanwhile, planning for OPTCON '91 in San José was moving ahead jerkily. The fundamental OPTCON model—a "core program" along with four technical meetings organized by each of the four societies—had inherent difficulties. Each society could have its own registration, and attendees wishing to hear papers in more than one society's meeting would have to pay for multiple registrations. This could become expensive, so OPTCON added the possibility of paying a single registration fee to register for all four meetings. But then there was squabbling among the societies about how to divide the revenue.

For OPTCON '91, the societies decided to allow an attendee to register for any one meeting, or to register for all four. And after complex negotiation, the societies agreed that OSA, SPIE, and LEOS would each get 29% of the joint-registration fees, and LIA, with a smaller program, would get 13%.

And then there was the issue of dividing the revenue from the exhibition among the five OPTCON sponsors. The original agreement for OPTCON '91 was that OSA, the exhibition manager, would receive twice as much as the other sponsors. That is, OSA would get two-sixths of the total, and the other four would get one-sixth. But in the spring of 1991, SPIE complained that was unfair because SPIE was also making a major contribution to the exhibition. In the end, it was agreed to divide the total revenue into seven parts, with OSA and SPIE each receiving two, and the other three receiving one.

Yet another point of contention was the subject matter of the societies' technical meetings. Some societies felt they had priority to address certain topics, and those topics should be off-limits for meetings organized by the other societies. This issue led to an hour's heated discussion at the May 16, 1991, steering committee meeting. By the time the discussion ground to a conclusion, it was decided that there were no rules; any society could organize sessions on any topic it chose.

By September 1991, the OPTCON scheduled for November of that year looked like it was being held together with rubber bands and thumbtacks. LEOMA was committed to sponsoring OPTCON '92, but the association's role after that was a prime topic at the September 26 meeting of LEOMA's executive committee. It was pretty clear that OPTCON was not having much effect on conference proliferation, but perhaps the conference really was beginning to attract the "new audience" of applications-oriented engineers. Some LEOMA members were ready to throw in the sponge, while others felt OPTCON still had promise. The executive committee decided to wait and see what happened at OPTCON '91 before making any decision.

But OPTCON '91 was something of a mixed bag. Income was up slightly from the OPTCON '88 in San José, but down considerably from OPTCON '90 in Boston. But the percentage of "walk-ins"—that is, attendees who simply walked into the exhibition without preregistering for the technical sessions—was up considerably from previous years. The OSA's Jarus Quinn observed that the "new audience" OPTCON sought was finally beginning to appear, because "walk-ins" to the exhibition are likely to be more interested in laser applications than attendees who register for the technical sessions (see Figure 2.3).

But OPTCON '91 did little to alleviate discord among the conference sponsors. Of the nearly 7000 attendees, a mere 179 had registered for the SPIE-organized component of OPTCON. Yaver explained that SPIE could

**FIGURE 2.3** Jarus Quinn, the longtime executive director of the Optical Society, was instrumental in organizing the OPTCON conferences.

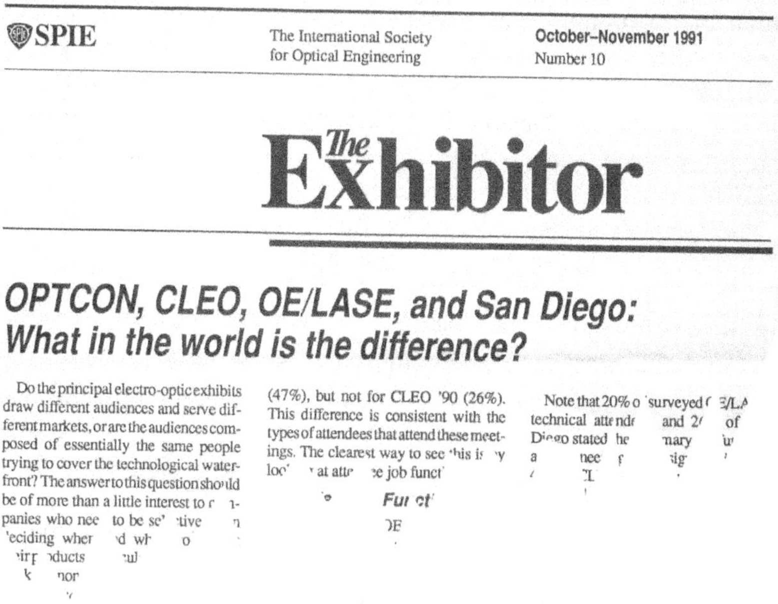

⊕ **SPIE**  The International Society  for Optical Engineering  **October–November 1991**  Number 10

# E**x**hibitor

## OPTCON, CLEO, OE/LASE, and San Diego: What in the world is the difference?

Do the principal electro-optic exhibits draw different audiences and serve different markets, or are the audiences composed of essentially the same people trying to cover the technological waterfront? The answer to this question should be of more than a little interest to companies who need to be selective in deciding where and wh... ...irp oducts ...

(47%), but not for CLEO '90 (26%). This difference is consistent with the types of attendees that attend these meetings. The clearest way to see this is by loo... at attr... e job funct...

*Fur ct*
)F

Note that 20% o surveyed 2/L technical attendr and 2 of Diego stated he nary a nee ...

**FIGURE 2.4** SPIE distributed a flyer at OPTCON '91 that claimed SPIE conferences attracted a larger applications-oriented audience than either OPTCON or CLEO.

not heavily promote a West Coast OPTCON to its membership because doing so would diminish participation at its existing West Coast conferences. Many more SPIE participants would be at OPTCON '92 in Boston, he pledged. Still, there was grumbling about SPIE's share of OPTCON revenue when their contribution had been so limited.

More resentment flared at the steering committee meeting when Ralph Jacobs, representing LEOS, distributed copies of a flyer that SPIE had distributed to OPTCON exhibitors in San José the night before (Figure 2.4). The flyer made a comparison among the major U.S. laser conferences, including CLEO, OPTCON, O-E Lase, and the SPIE Annual Meeting in San Diego. The latter two conferences—both SPIE products—attracted far larger "engineering and design audiences" than either OPTCON or CLEO, according to the flyer. Distribution of the flyer at all, but especially at the OPTCON conference, was grossly inappropriate, Jacobs asserted. Yaver replied that it was merely an innocent attempt to provide exhibitors with helpful information. Sprague, representing the SPIE board, saw the need for greater diplomacy and apologized for the distribution at OPTCON.

Assuming the steering committee chair in 1992, the LIA's Murray Penney took charge of a committee even more agitated than the one Silfvast had inherited the year before. The OPTCON marketing committee, chaired by LEOMA's David Rossi, tabled a proposal at the May 1992 meeting that offered the potential of calmer waters. Rossi proposed that the marketing committee prepare a long-range strategic marketing plan that would guide OPTCON from a technical conference with an attached equipment exhibition, to a stable applications-oriented trade show with few if any purely technical sessions. There would be four phases to the plan:

1. Identify those industries that would benefit from increased use of lasers.
2. Determine what information those industries need.
3. Modify OPTCON to be a vehicle to convey that information to those industries.
4. Market OPTCON as an invaluable tool to those industries.

Designing and implementing the plan would cost $33,500, Rossi's committee estimated.

The steering committee unanimously approved Rossi's suggestion, and agreed to devote $7,000 from the OPTCON '92 budget to starting work, with the balance to be included as a line item in the OPTCON '93 budget.

In the spring of 1992, planning for OPTCON '92 was moving ahead based on the existing model of a core program surrounded by technical conferences organized by each of the four technical societies. Agreements about distribution of conference revenue, and about which topics would be addressed by which societies, were reached in off-line discussions among the various sponsors. In April, SPIE—the conference manager for OPTCON '92—announced that it would continue to be an OPTCON sponsor after 1992, ending an uncertainty that had existed since the January 1991 steering committee meeting.

But the division within LEOMA about OPTCON was becoming more pronounced. Spectra-Physics and Coherent, two of LEOMA's largest members, decided not to exhibit at OPTCON '92: Their reasoning was that OPTCON was doing nothing to reduce conference proliferation, and the "new audience" simply was not materializing. At the same time, other LEOMA members supported Rossi's plan to convert OPTCON into an equipment exhibition targeting certain industries.

Because several of the larger exhibitors at previous OPTCONs were not participating in 1992, OPTCON '92 booked a smaller exhibition space at Boston's Hynes Auditorium than it had in 1990. By the time the conference took place, the space was completely sold out, but revenue was considerably less than it had been at previous OPTCONs (Table 2.1).

TABLE 2.1   **Diminishing Returns: Net Revenue from OPTCON by Year**

| 1988 Santa Clara | 1990 Boston | 1991 San Jose | 1992 Boston |
|---|---|---|---|
| $326,000 | $856,000 | $341,000 | $72,000 |

And while surveys of the exhibitors at OPTCON '92 were inconclusive, there was little evidence that applications-oriented engineers were appearing in greater numbers than at previous OPTCONs. At a board meeting held during OPTCON '92, LEOMA elected to end its financial participation in future OPTCONs—LEOMA would have a share in neither the profits or the losses. But still hoping to draw the ever-elusive "new audience," LEOMA agreed to continue supporting OPTCON by organizing a dozen applications-oriented sessions at OPTCON '93 in San José.

At the January 1993 meeting of the OPTCON steering committee, Rossi presented the strategic marketing plan that his committee had devised during the previous seven months. It was a professionally designed marketing plan, describing 20 industries—for example, biomedical instrumentation, environmental monitoring, and the military—that would benefit from increased utilization of lasers and photonics. And the plan laid out a timeline explaining how OPTCON could change to address those industries during the next four years. But in extensive interviews with existing OPTCON exhibitors, the marketing committee discovered that exhibitors had a strong desire to see both the new applications-oriented audience and the traditional research audience, at OPTCON. Therefore, in order to maintain a strong exhibition, the committee recommended that OPTCON retain the technical sessions organized by professional societies. That recommendation was a disappointment to many, who had hoped that OPTCON would transform into something completely different.

Nonetheless, the steering committee in January unanimously approved moving ahead with OPTCON '93 in November in San José, as outlined in Rossi's timeline. That called for a continuation of the core program with short courses, how-to seminars, an employment center, and product presentations. But only one professional society, LEOS, would organize a technical session at OPTCON '93.

But the enthusiasm in the steering committee didn't extend to the community at large. By the summer of 1993, it was becoming clear to even the most enthusiastic supporters that OPTCON was on its last legs. "There was a meeting of the minds," Sue Davis recalled in a recent conversation. "[OPTCON] just wasn't financially viable." Davis, a ranking executive at SPIE, was responsible for selling exhibit space at many of the OPTCONs, and saw first-hand the flagging interest that exhibitors had in OPTCON.

**FIGURE 2.5**   Joe Yaver, the executive director of SPIE, maintained that SPIE and OSA had vastly different philosophies on managing technical conferences.

At its board meeting held during OPTCON '93, LEOMA elected to withdraw completely as an OPTCON sponsor. OPTCON '93 was the last OPTCON.

"Internal conflicts killed it," SPIE's Yaver told me in 2012. "The cultures [of the OPTCON sponsors] weren't compatible. OSA likes to do things in committees. Joe Yaver likes to get a few people together over a beer and decide." (see Figure 2.5). That statement, in a nutshell, summarizes the attitudes that have historically hampered the photonics community from unifying politically, and perhaps continue to do so today.

## CPO

Several years later, the seeds for another joint venture within the photonics community began to germinate. In early 1995, the National Research Council established a committee to prepare a report on the status of optics technology in the United States. The report was to evaluate the current and future impact of optical science and engineering on the U.S. society, and to prioritize the technical opportunities in the context of national needs. The Committee on

Optical Science and Engineering (COSE, pronounced "cozy") set out on a series of workshops across the country to fulfill its task.

LEOMA and representatives of the laser/optics professional societies met with COSE in March 1996 to provide their inputs for the report. Speaking for LEOMA, I listed the four major challenges facing the U.S. laser/photonics industry. Market competition between government agencies and the private sector led the list.[3] Close behind was excessive government regulation, especially of export controls, and foreign control of international standards.[4] The fourth area of concern was the excessive cost of intra-industry litigation.[5]

That March 1996 workshop sparked the interest of several individuals in the photonics community. In particular, the leadership of the OSA and SPIE recognized that the COSE Report could focus federal attention on the issues of concern to the community, and facilitate their solution. Working together, the two professional societies organized a meeting at the 1996 CLEO conference to examine how the community could enhance the impact of the COSE Report when it was released.

Invitees to that meeting included the professional societies and trade associations active in optics and photonics:

- APOMA, the American Precision Optics Manufacturers' Association
- LEOS, the Laser and Electro-Optics Society
- LEOMA
- LIA, the Laser Institute of America
- OIDA, the Opto-electronics Industry Development Association
- OSA, the Optical Society of America
- SID, the Society for Information Display

There was general agreement that the entire photonics community stood to benefit from the COSE Report, but in the summer of 1996, nobody except perhaps the COSE members themselves knew what the report was going to say. In particular, Arpad Bergh, the executive director of OIDA, cautioned against encouraging widespread dissemination of the report before its contents were known. On the other hand, advocates argued that waiting until the report was released would sacrifice the ability to act in a timely manner when the report was released. Don Scifres, the 1996 LEOMA president, and I

---

[3]This issue was the subject of congressional testimony I had given several years earlier, as described in Chapter 6.

[4]LEOMA's accomplishments in export controls are described in Chapter 5, and in international standards in Chapter 3.

[5]LEOMA's Alternative Dispute–Resolution Agreement, described in Chapter 7, reduced intra-industry litigation.

represented LEOMA at the meeting, and we joined the consensus in favor of moving quickly to support the COSE Report, even if its contents were unknown.

A loose coalition, initially called the Coalition for Optics (CFO) was organized at the CLEO meeting. The coalition would not incorporate, and it would not handle any funds; it was to be nothing more than an informal coalition of professional societies and trade associations. Duncan Moore, the 1996 OSA president, was elected the first chair.

But at that initial meeting, another concern came to the fore. "Electrical engineering" was a profession, "physics" was a profession, "mechanical engineering" was a profession, but "photonics" or "optical engineering" was not broadly recognized as a profession. Instead, schools that offered advanced optics courses, for the most part, offered them as part of the electrical engineering curriculum or as part of the physics curriculum. Compensation surveys were available for fields like electrical and mechanical engineering, but aside from the limited LEOMA survey,[6] there were no compensation data available for optical engineers. In general, the lack of broad recognition for "optical engineering" as a discipline made it difficult to find employees, to compensate employees, and to conduct business in general.

So, by the end of that first CFO meeting, two goals for the coalition had been identified: publicize the COSE Report, and promote "photonics" as a professional discipline.

And, later during 1996, the CFO participants sought other projects that the coalition could address more efficiently than they could individually. LEOMA agitated for broader community support of its activities in international laser standards, and there was a recognized need for better training at all levels, from high schools through graduate schools. But still, the emphasis was on the COSE Report, whose release was expected within the year. Looking ahead to the political task of promoting the report, Gary Bjorklund, representing OSA, proposed creating an "optical census." The concept was to survey the entire United States, seeking to quantify the economic contribution of the photonics industry to each congressional district.

As more societies and associations joined the CFO during the latter half on 1996,[7] several voiced concern over the name of the coalition. "Optics" to some people was too limiting a word. In that view, the word included only classical optics dealing with lenses, rays, waves, and the like, but did not extend to quantum laser physics or many other topics of interest to CFO participants. In

[6]The LEOMA compensation survey, begun in 1996, is described in Chapter 7.

[7]By the end of 1996, CFO had eight members: the Arizona Optics Industry Association (AOIA), APOMA, the Society for Imaging Science & Technology (IS&T), LEOMA, LEOS, LIA, OSA, and SPIE. Also, the International Commission for Optics joined as an observer.

November 1996, the name was changed to Coalition for Photonics and Optics (CPO) to reflect the larger purview of its members.

Although a draft of the table of contents of the COSE Report was circulated in late 1996, the content and recommendations of the report remained a mystery. CPO requested a prepublication review of the report, with an eye toward underwriting the cost of publication. The request was politely declined.

The COSE Report—officially titled *Harnessing Light: Optical Science and Engineering for the 21st Century*—was eventually published in August 1998. It explicitly addressed six fields in which optics would play a crucial role: information technology; health care; energy and lighting; national defense; manufacturing; and research and education. The report also dedicated a chapter to the issue of manufacturing optical systems.

The report found that "Optics is a critical enabler of technology in many industries," and that "Optics will help revolutionize communication, medicine, energy efficiency, defense and manufacturing." It recommended increased optics funding by numerous federal agencies, including the National Institutes of Health and the Defense Advanced Research Projects Agency.

These were, of course, the kinds of recommendation the community could support. CPO and its members immediately set about publicizing the report. CPO itself issued a blanket endorsement of the report. Vu-graphs and PowerPoint presentations were prepared to help individual members explain the report to supervisors and funding sources. LEOMA prepared a white paper urging NIST to take COSE's advice and provide financial support for the country's activities in optics standards. As described in detail in Chapter 7, Randy Heyler, LEOMA's 1998 president, led several dozen industry executives—mostly CEOs—to Washington where they met with their respective congressional delegates and discussed the COSE Report. There was a LEOMA executive seminar in Washington to which we invited several key congressional staffers to attend and share their perspectives on the COSE Report.

The U.S. Senate had created a Manufacturing Task Force to identify steps needed to strengthen domestic manufacturing. CPO launched an intense lobbying campaign to bring the COSE Report to the attention of that task force. Congressman Vern Ehlers—at that time the only physicist in Congress—had conducted an independent study of science and technology in the United States, and many of his conclusions paralleled those of the COSE Report. In congressional visits, CPO members sought to link the two studies and thereby strengthen the impact of both.

But, looking back, it's hard to see that the COSE Report had much impact on funding or on resetting anybody's priorities. It did, according to COSE member Bob Byer, provide a valuable "snapshot in time" of the state of the technology in 1998. But its impact on funding levels was "somewhat subtle," according to Elsa Garmire, another COSE member.

"I think . . . it is very difficult to measure the direct impact of such a report. Often reports are one of many inputs that sway the national discourse," Chuck Shank, the chair of COSE, told me recently. Ironically, the report probably had more effect on optics funding in Europe than in the United States, according to Tom Baer, who consulted with COSE in the preparation of the report.

For the LEOMA companies that invested thousands of dollars worth of their executives' time disseminating and publicizing the COSE Report, the return on that investment was scant to nonexistent. The NRC published a follow-up to the COSE Report in 2012. The funding for the follow-up was a fraction of the funding for the original report, and the laser companies that were so heavily involved in promoting the original report pretty much ignored the follow-up.

What became of the other projects CPO had identified? The "optical census," which had been suggested by Gary Bjorklund, was a fine concept, but it turned out to be difficult to implement. The idea was to build political strength for the photonics community by quantifying the jobs and revenue that optics—"optics" in the broadest sense—created in the United States. Perhaps the numbers could even be broken out by congressional district. Initially, there were reports that such data already existed within the Department of Commerce. Eventually it turned out they did not. Then there were concerns about privacy and confidentiality. And the sheer magnitude of the census was overwhelming for a group of volunteers. The project remained on the CPO agenda for years until 2000, the year I chaired CPO, when I asked the members to decide if they wanted to undertake the project or abandon it. With little discussion, CPO decided to abandon it.

Identifying "photonics" as a full-fledged discipline likewise turned out to be a difficult concept to implement. A group of LEOMA executives visited the Bureau of Labor Statistics (BLS) in 1998, urging them to identify "photonics" as an employment category in their statistics. But they told us BLS categories were based on usage: help-wanted advertising, company classifications, and so forth. If we wanted "photonics" to be a category at BLS, we should advertise for "photonics engineers" and lobby government labs to classify people as "photonics technicians." In recent years, the word "photonics" has perhaps become more common—one or two universities have established photonics departments—but CPO can claim little credit for these changes.

And, despite LEOMA's best efforts, CPO never embraced international photonics standards. As explained in more detail in Chapter 3, the community's attitude seemed to be that standards were the responsibility of the for-profit private sector, not of the broader academic photonics community.

TABLE 2.2   Chairmen of the Coalition for Photonics and Optics (CPO)

| 1996 | 1997 | 1998 | 1999 | 2000 | 2001 | 2002 | 2003 |
|------|------|------|------|------|------|------|------|
| Duncan Moore OSA | Duncan Moore OSA | Randy Heyler LEOMA | Gary Bjorklund OSA | Breck Hitz LEOMA | Gordon Day LEOS | Don O'Shea SPIE | Bob Breaux AOIA |

But Chapter 4 explains that, by the year 2000, the photonics community was facing a shortage of skilled employees that threatened to undermine the industry's rapid growth. The market for components of fiber-optic communication systems, in particular, was growing explosively. "You just walk into the room with a basket of [fiber-optic components], and instantly you have a crowd of customers around you," Spectra-Physics President Pat Edsell commented at the time. Virtually every photonics company in the United States was desperate to hire skilled—or even unskilled—workers, but there were no workers available.

The reason for the shortfall was readily evident in data from the Department of Education. Those data showed, among other things, that the number of degrees in electrical engineering awarded by U.S. schools had been dramatically decreasing in recent years, to the point where—as LEOMA director Dave Hardwick liked to say—the schools were graduating more gym teachers than electrical engineers (Figure 2.6). In the short term, fiber-optics manufacturers faced an immediate limitation on their growth, but in the longer term the situation threatened (and still threatens) the country's technical competitiveness and national security.

LEOMA had already started to address the problem on its own, but as incoming chair of CPO in 2000 (see Table 2.2), I proposed that the coalition refocus its efforts away from the COSE Report and other projects, and onto the issue of worker training. The proposal was favorably received, and many of the individual CPO members had already begun programs to enhance the nation's educational programs in optics and photonics. There was unanimous agreement to modify the CPO mission statement to include "Projects to facilitate a greater awareness of careers in optics and photonics." As a first step, staff members of several societies set about designing a CPO website that would offer information about careers in photonics. CPO became the instrument to encourage and coordinate members' various projects promoting photonics education.

Chapter 7 describes LEOMA's annual executive seminars, which by 2000 had been taking place in Washington, D.C., for a number of years. But the 2000 executive seminar was to be special: titled "So many jobs . . . So few people," it was cosponsored by the CPO and focused on the issues of education and training in the field of photonics. Speakers included Duncan Moore, who had been one of the instigators of CPO, and who in 2000 was a

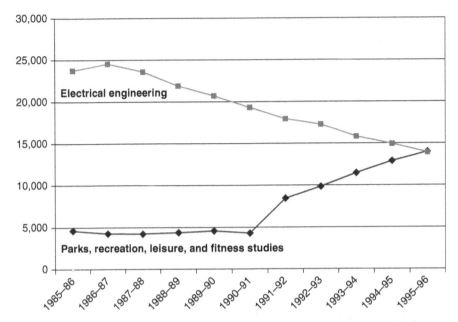

**FIGURE 2.6**   According to data from the Department of Education, U.S. schools were granting more undergraduate degrees in recreational studies than in electrical engineering.

deputy director of the White House Office of Science and Technology Policy; Cathleen Barton of the Semiconductor Industry Association, who explained how that industry was coping with worker shortages; and Cosette Ryan of the Education Department, who described that departments accelerated graduate assistantship programs in technology fields.

LEOMA's 2000 president, Bob Phillippy of the Newport Corporation, also spoke, describing the PowerPoint presentation he and others at LEOMA were designing, aimed at high school seniors and college freshmen, and extolling the benefits of careers in optics and photonics. He explained that each member of the LEOMA board would make the presentation to at least two groups every year, and asked how many attendees at the seminar would agree to use the presentation at high schools and colleges in their areas. Several dozen responded positively.

Aimee Gibbons of the OSA also spoke at the seminar, describing an "optics suitcase" that had been designed at the University of Rochester. The suitcase contained equipment for performing intriguing demonstrations of optics principles to high school students, and the plan was to distribute the suitcase widely to high schools in the United States.

Under the leadership of Gordon Day in 2001, CPO continued to focus its energy on education and training. CPO members pledged thousands of dollars to underwrite the mass production and distribution of the Phillippy/LEOMA

PowerPoint presentation for high school seniors and college freshman. Penn State was prepared to underwrite much if not all of the cost of manufacturing and distributing the OSA optics suitcase.

But by late 2001, it was becoming clear that the issue of *So many jobs, So few people* was beginning to reverse itself. The booming market for fiber-optic components was collapsing. The technicians and engineers who had so recently been hired by desperate photonics companies were being laid off. The severe shortage of skilled workers that CPO had been striving to alleviate suddenly and spontaneously disappeared. The PowerPoint presentation that Bob Phillippy and others had labored long and hard to complete was presented less than half a dozen times. The OSA optics suitcase was distributed, but has never achieved the utilization visualized in 2000.

So CPO ended in 2001, no longer addressing any major mission. Created to publicize the COSE Report, CPO had shifted its attention to education and training in 2000, but now that issue had become moot. CPO appointed a committee consisting of its incoming, 2002 chair, Don O'Shea and the two previous chairs—Gordon Day and myself—to develop a new strategic plan.

Absent a pressing need, was there really a need for the existence of CPO? That was the question the committee presented to the CPO members at the first meeting in 2002. Pressing need or not, there was a reluctance to abandon the concept, and the CPO membership voted overwhelmingly that CPO should continue.

Meanwhile, Duncan Moore, having completed his term at the Office of Science and Technology Policy, brought some of his experience back to the photonics community. The National Nanotechnology Initiative (NNI) had been very successful at advancing that technology, and Moore proposed that the photonics community undertake a similar effort. Two of the leading professional societies, OSA and SPIE, agreed. They jointly hired Brendan Plapp, who held a PhD in physics from Cornell and had recently completed a congressional fellowship sponsored by the American Physical Society to coordinate the effort. Plapp managed the National Optics and Photonics Initiative (NOPI), a political effort to increase federal funding for photonics. At the request of OSA and SPIE, CPO identified support of NOPI as a major component of its new mission.

The hope was that the Office of Science and Technology Policy would support NOPI the same way it had supported the NNI. But after November 11, 2001, federal agencies were much more concerned with national security than basic science, and that support never materialized.

This disappointment dampened the enthusiasm of CPO's membership to continue. In early 2003, SPIE and OSA wrote a joint letter to the incoming CPO chair, Bob Breault, questioning whether their investment of staff and volunteer time brought sufficient return to justify continuing the organization.

Breault argued that CPO served a valuable function as a communication hub among the various societies and associations that were its members, but it was not clear that the members agreed. SPIE announced that it would no longer provide a staff member to take minutes at the meeting. More seriously, nobody was willing to assume the chair of CPO in 2004.

The last CPO meeting took place in June 2003. There was discussion of refashioning NOPI into an organization that would generate roadmaps for the development of optics and photonics in the United States, but the enthusiasm just was not there. I reported that the LEOMA board felt roadmapping was too nebulous an undertaking, and the focus should be on more immediate concerns like export controls and international standards. Paul Shumate explained that the LEOS board had discussed roadmapping, but decided they could not support the project financially. With no member willing to provide secretarial services for the meeting's minutes, and with no candidate to chair CPO in 2004, CPO came to an end.

### The Benefit of Hindsight

Neither of the two efforts to bring the photonics community together for the common good was successful. LEOMA had stimulated the creation of OPTCON to reduce conference proliferation, and the conference's secondary goal—attracting a "new audience"—was something of an afterthought. But OPTCON succeeded at neither goal.

Reducing conference proliferation really was not in the societies' best interest. Conferences are a source of significant revenue for the professional societies that sponsor them, and OPTCON drew too small an audience to replace the revenue the societies normally gained from their stand-alone conferences. In the years following OPTCON's demise, the issue of conference proliferation pretty much went away. Coherent and Spectra-Physics had broken new ground by withdrawing their exhibitions from a major conference—OPTCON—and thus became less reluctant to be absent from other conferences. Other manufacturers followed suit, and eventually the community settled into having one or two major North American conferences each year, and a host of secondary, specialized conferences with limited equipment exhibitions.

OPTCON's secondary goal, attracting a new, diverse audience of applications-oriented engineers who didn't attend the existing laser conferences, was probably never realistic. As Milton Chang predicted during OPTCON's earliest years, once an application becomes successful enough to draw a significant audience, a conference for that application alone will be created.

The Coalition for Photonics and Optics, the community's second attempt to come together, met the same fate that LEOMA ultimately met: it was a solution in search of a problem. Created to address two issues, publicizing the COSE Report and enhancing the recognition of the field of photonics, it had limited success with both of them. Then, in 2000, CPO identified the acute shortage of trained photonics workers as a problem indeed worthy of its attention—only to have that problem vaporize. Lacking a viable problem to address, CPO ceased to exist.

# 3

# INTERNATIONAL LASER STANDARDS

If export controls and "conference proliferation" were the seminal issues behind the creation of a trade association for the North American laser industry, international standards quickly became at least equally important. As early as January 1987, the LAA board initiated its involvement with the laser committee of the International Electrotechnical Commission (IEC). Jerry Glen, the technical director of the Laser Institute of America (LIA), was serving as secretariat of the IEC Laser Committee, and the LAA board agreed to cosponsor that activity with the LIA; each organization designated $6500 for this purpose in 1987. The IEC Laser Committee, officially IEC Technical Committee 76 (IEC TC76), was addressing issues of eye safety around lasers.

In July 1987, Glen attended the IEC TC76 meeting in Prague, and reported back on the progress at that meeting. The LIA and LAA both found the report reassuring. There were discrepancies between the standards being developed in IEC TC76 and those promulgated by the U.S. Center for Devices and Radiological Health (CDRH), a division of the FDA. It was important to eliminate these discrepancies, lest U.S. manufacturers be required to produce different lasers for domestic and international customers. But the members of IEC TC76 were aware of those differences, and were apparently working to eliminate them.

Supporting Glen was an expensive undertaking, and the question arose whether it was worth doing again in 1988. After all, everything seemed to be

*LEOMA and the U.S. Laser Industry: The Good and Bad Moves for Trade Associations in Emerging High-Tech Industries*, First Edition. C. Breck Hitz.
© 2015 by The Institute of Electrical and Electronics Engineers, Inc. Published 2015 by John Wiley & Sons, Inc.

going smoothly in IEC TC76. I was assigned to look into the matter and make a recommendation to the board on whether the support should continue for another year.

I recommended that LAA continue cosponsoring Glen's work with IEC TC76 in spite of the significant expense. There were numerous incidents in which international standards had been used to provide a competitive advantage to one nation or another, and one example was particularly relevant to U.S. optics manufacturers. In the early 1980s, Germany adopted a standard for the threads on microscope objectives, and proposed that the standard be adopted by the International Organization for Standards. The standard, which was defeated only at the last moment by U.S. objections, would have required U.S. manufacturers like American Optical and Bausch & Lomb to spend millions of dollars to reengineer their microscopes and lenses. As secretariat of IEC TC76, Glen saw every standard the laser committee addressed, and could sound the alarm if there were any attempts to introduce unfair standards. The LAA board accepted my recommendation.

And the board thought that the thousands of dollars a year it was spending to underwrite Glen's work would solve the issue of international laser standards. When that turned out not to be the case, the initial reaction was one of annoyance. As the European Union was coming together—"Europe '92" was the phrase widely heard—there was talk of new European standards affecting all sorts of U.S. imports to that continent. Would these standards absorb even more resources from U.S. laser manufacturers?

Despite the hope that we had eliminated international standards as a problem by supporting work in IEC TC76, the answer to that question appeared to be "yes." In January 1989, the *Wall Street Journal* reported[1] that new European standards would be incompatible with existing practice of U.S. forklift manufacturers, making U.S. forklifts more expensive and less competitive in Europe. "Many U.S. companies will face similar problems as Western Europe re-regulates its way to economic unity," the article predicted. Clayton Yeutter, then the U.S. Trade Representative, warned that "The implications of 1992 for the U.S. are far more serious than most people realize."

"Even American [companies] that are established and think they are grand-fathered, are at risk," echoed Charles Ludolph, the Commerce Department's director of European Affairs. And, striking closer to home, we learned that the wavelengths specified in proposed European standards for measurements of optical glasses were different from those used in the United States. If adopted,

---

[1]The article was written by Walter Mossberg, who 30 years later would be well known as the newspaper's computer guru.

**TABLE 3.1    The European Standard Committees Reflect the Charters of the International Committees**

| International | European | |
|---|---|---|
| IEC | CENELEC | Standardizes electrical equipment |
| ISO | CEN | Standardizes everything else |

the new standards would require a vast and expensive recalibration of the U.S. instrumentation.

The administrative nature of international and European standards can be confusing, and is summarized in Table 3.1. In 1989, there were two European standards organizations, the European Committee for Electrotechnical Standardization (CENELEC) and the European Committee for Standardization (CEN). These two committees reflected the charters of the two international committees chartered by the United Nations, the IEC, and the International Organization for Standards (ISO).[2] The IEC's charter is to standardize electrical devices, while the ISO creates all other standards.

It was the IEC Laser Committee (IEC TC76) that was developing standards for laser eye safety. But there were no standards for the lasers' parameters themselves—power, stability, polarization, and so forth. In order to implement their "integrated market," the Europeans needed standards—on literally everything—that would facilitate intra-European trade. And the CEN Laser Committee (officially CEN TC123) was in a headlong rush to create performance standards for lasers before the 1992 deadline.

It was these CEN laser standards that threatened the European sales of LAA members. The LAA Board initially contemplated getting the work transferred from CEN TC123 to the IEC TC76, where Jerry Glen could exert some control over the work. It turned out, however, that that was not a viable option. The Europeans had already assembled their laser experts into CEN TC123, and they were trying to get their new standards approved by 1992. They would not even consider transferring the work to IEC TC76.

In March 1989, I was contacted by Sidney Braginsky, an executive at the U.S. division of Olympus Corporation who had been active in the ISO Optics Committee (Technical Committee 172, or ISO TC172). He explained that ISO was in the process of creating a laser committee that would interact with CEN TC123. The first meeting of the ISO Laser Committee—officially a subcommittee of the ISO Optics Committee, and designated as ISO TC172 SC9,

---

[2]As explained in Chapter 1, it is incorrect to take ISO as an acronym, and call the organization the "International Standards Organization." Instead, "ISO" is Greek for "same" or "equivalent." The goal of standardization is to make measurements, procedures, and so on the same everywhere they are performed.

or simply "SC9"—was scheduled for later that spring in Germany. Braginsky had been appointed chairman of that committee, and he urged LAA to send delegates to participate in its meeting.

If we in the United States were confused by which committee was supposed to create which laser standards, the Europeans were apparently equally confused. The creation of a laser committee within ISO set off an alarm at the highest levels of IEC, where the ISO Laser Committee was viewed as an intrusion into territory already claimed by IEC TC76. Not so, replied the ISO leadership, which shares office space in Geneva with the IEC. Lasers are optical devices, and fall well within the jurisdiction of the ISO Optics Committee. It was a disagreement that would take the better part of two years, and at least two high-level meetings in Geneva, to settle.

With the ISO versus IEC feud simmering in the background, the first meeting of the ISO Laser Committee—SC9—was postponed until an unspecified date in fall of 1989. But work on laser standards in CEN TC123 continued unabated. In the late summer of 1989, Bob Weiner, a U.S. expert on laser safety standards and a member of IEC TC76, provided me with a copy of some of the proposed CEN laser standards. They were frightening! They specified many parameters, from the sizes of laser mirrors to the position of mounting holes, that were different than accepted dimensions of U.S. manufacturers.

The LEOMA[3] companies deemed the proposed standards "very threatening" when I circulated the documents to them. At a board meeting in late August, Jon Tompkins, the Spectra-Physics CEO, proclaimed that "many people around this table will be out of business in Europe" if the proposed standards were enacted. Tompkins suggested that a special fund be established, drawing tens of thousands of dollars from each of the larger LEOMA members, and used to underwrite LEOMA's response to the European proposals.

The board agreed that drastic steps were called for, but wanted a better understanding of the situation before deciding what steps would be appropriate. I was directed to request experts at each of the companies represented on the board to prepare a one-page appraisal of the proposed CEN standards. I would then compile those pages into a single document and preface it with an executive summary of the experts' appraisals. That way, the impact of a two-inch-thick stack of proposed standards could be summarized in a 15-page document.

That document induced the board, at its next meeting in October 1989, to adopt Tompkins's suggestion of a special fund dedicated to international standards—even though it was not clear how that money should be used. The

[3]The organization changed its name from LAA to LEOMA in June 1989 (see Chapter 1).

board agreed that each LEOMA member company would be asked to donate twice its annual dues to the fund. The monies so raised would be dedicated to generating alternatives to the CEN standards that would be less objectionable to U.S. manufacturers, and to figuring out the politics of getting those standards substituted for the existing CEN proposals.

The first step was by far the simpler of the two. We hired a consultant, Tom Lieb, to begin drafting alternatives to the CEN standards. The second step was more problematic. How could we get our revised standards into the European system—or even get the Europeans to acknowledge the existence of the U.S. position? There were at least three options. We could take our new standards to what was then the National Association of Photographic Manufacturers (NAPM), a U.S.-based organization for optics-related standards. NAPM would adopt the standards and forward them to international standards organizations, where they could become globally recognized. Faced with worldwide recognition of the NAPM standards, CEN would be forced to withdraw its standards. This option appeared to be the least direct of the three.

The second option was to take our standards to the IEC, to IEC TC76, where Jerry Glen was the secretariat and many U.S. experts already participated. IEC would adopt the standards as fully authorized IEC International Standards, and worldwide recognition would be automatic. Again, faced with globally recognized standards, CEN would be compelled to withdraw its Europe-only standards. This option was attractive because it would build upon LEOMA's existing support of IEC TC76. The drawback was that the withdrawal of the CEN standards could not be certain, and the conflict was likely to be ugly until it was resolved.

The third option was to take our new standards to the newly formed ISO Laser Committee, known simply as SC9. If ISO created acceptable standards for lasers, CEN was obligated by its own rules to abandon its standards and adopt the ISO standards. Sid Braginsky, the newly appointed chairman of SC9, advocated this approach. It was attractive in that it avoided the potentially ugly conflict with CEN standards. We would be working with CEN 123 rather than against it. The drawback was that we had zero understanding of the dynamics of working with a new ISO committee. We understood how to work with IEC TC76, and were confident we could enact reasonable standards there. But the new SC9 would include—and maybe be dominated by—many of the same European engineers who participated in CEN 123. There was no way of knowing whether our revised standards could be accepted there, or if the European engineers would insist on the same standards they had created in CEN 123.

Jerry Glen, with whom we had worked for several years and whom we trusted, strongly advocated the second option. He passed on to us assurances from the IEC leadership in Geneva that IEC would actively support IEC laser standards in the face of CEN opposition. Sid Braginsky, on the other hand, passed on assurances from the ISO leadership and from Hilmar Ganser, secretariat of ISO TC172, that the European engineers in SC9 would be eager to cooperate with U.S. participants in SC9. Ganser's pledge carried significant weight, because he was also secretariat of CEN TC123.

LEOMA had organized a Technical Advisory Group (TAG) of U.S. experts working with Tom Lieb to fine-tune the laser standards the United States would propose—although it still was not clear to whom they would be proposed. In November, Karlhanns Gindele, who had been designated by ISO to eventually replace Hilmar Ganser as the secretariat of SC9, joined a TAG meeting at my invitation. It was my first meeting with Gindele, a seemingly straightforward German fellow whom I instinctively liked. Gindele reiterated the message we'd gotten from Ganser, that the European engineers from CEN TC123 were anxious to work with their U.S. counterparts to forge laser standards that would be as useful to Americans as to Europeans. The draft CEN standards—the ones that had caused such consternation in the United States—were the best the European engineers could do without input from the United States, Gindele insisted. But they were anxious to work with U.S. experts to make them more palatable to U.S. manufacturers.

What we faced was the age-old question of the cart and the horse—Which comes first? In principle, the first decision would come from the upper levels of ISO/IEC management in Geneva: Which organization, ISO or IEC, would be responsible for standardizing lasers? Then LEOMA would decide whether it would work with the ISO Laser Committee (SC9) or the IEC Laser Committee (TC76).

In practice, both committees knew that LEOMA's preference could make a big difference in Geneva, because the U.S. laser industry was by far the largest in the world. If the U.S. laser industry cast its lot with the new ISO Laser Committee, then ISO could argue in Geneva that it had support of the majority of global laser manufacturers, and the IEC laser-safety standardization should be turned over to ISO. If LEOMA sided with IEC, IEC could make a convincing argument that the laser-parameter work that was beginning in ISO be turned over to IEC.

So both TC76 and SC9 lobbied for LEOMA's support. When he visited the TAG meeting in November 1989, Gindele informed us that the initial meeting of SC9 would be moved from Germany to the United States, "to facilitate American participation."

But the LEOMA board felt it lacked sufficient understanding of the politics and dynamics underlying the ISO/IEC issue to make an intelligent decision. To gain a better understanding, the board decided we needed a European consultant and/or attorney, and I was dispatched to Brussels to try to find a suitable party. I interviewed several firms in Brussels, and selected Clifford

Chance, a firm specializing in European Union issues, to advise LEOMA about E.U. issues in general, and standardization in particular.

On that same European trip in December 1989, I attended a meeting of upper management at the ISO/IEC offices in Geneva. The meeting had been called in an attempt to settle the conflict between ISO and IEC over laser standardization. It was jointly chaired by Roger Desmons, the IEC's director of External Communications, and Ch. Favre, the ISO's deputy secretary general. Included among the attendees were Tony Rayburn, the secretary general of IEC; Bengt Kleman, the chairman of IEC TC76; Karlhanns Gindele, designated to become the chairman of SC9; and other ISO and IEC officials. I was the only non-European in the room.

I summarized the U.S. perspective: We now felt that laser standardization should be divided between TC76 and SC9. TC76 had the expertise and experience to craft good standards for laser eye safety, but not for standardization of laser parameters like power, wavelength, and beam propagation. CEN TC123 was already creating European standards for these and related parameters, and SC9 had a better interface with that CEN committee than IEC TC76 did. Therefore, the standardization of laser parameters should be done in SC9.

The ISO/IEC management basically accepted LEOMA's position, and the meeting subsequently set about the task of explicitly defining which laser parameters SC9 would standardize. Those clearly included all basic laser operating parameters, but also added to SC9's list were laser components— mirrors, polarizers, beam splitters, and so forth—as well as marking and labeling of lasers. But this meeting still left unresolved many other details about the division of responsibility between ISO and IEC. It wasn't clear, for example, which organization would generate standards for the emerging field of fiber optics, or how the different aspects of safety standardization would be divided.

Even as the Geneva meeting was taking place, ISO announced that the inaugural meeting of SC9 would take place in San Francisco in March 1990. LEOMA supported that meeting, and we recruited more than two dozen senior laser engineers from member companies to participate in it. These engineers spent copious time studying the draft CEN standards and working with Tom Lieb and others on the U.S. revisions.

By January 1990, the voluntary standards assessment launched at a LEOMA board meeting in October had collected over $140,000 from LEOMA companies. The board directed me to earmark about $70,000 to underwrite Tom Lieb's project to revise the draft CEN standards, and another $20,000 to pay for Clifford Chance, the Brussels attorney. The remaining funds were left to my discretion for travel and other standards-related expenses.

In February I was in Europe again, this time to visit the offices of Karlhanns Gindele and Hilmar Ganser in Pforzheim, Germany. There I met Adolf Giesen, a leading member of CEN TC123, and Peter Greve, the chairman of TC123. They were both eager to participate in the new SC9. All these gentlemen were well acquainted with the choreographed management of international standards meetings, and showed me the agenda that they had carefully prepared for the upcoming SC9 meeting. As I knew, TC 123 was in a rush to complete laser standards before the unification at the end of 1992, and could not wait for SC9 to move at the usual pace of ISO committees. If SC9 was to influence the European laser standards, we would have to have our own standards ready by 1992. We agreed that, to do so, SC9 would have to meet twice a year in 1990 and 1991, and all the participating engineers would have to devote a significant amount of time to the projects between meetings.

LEOMA's first executive seminar[4] was held in Monterey, California, the weekend immediately before the inaugural SC9 meeting in San Francisco, and I invited prominent participants in the SC9 meeting to join the industry executives at the seminar. Sid Braginsky, the chairman of SC9, Karlhanns Gindele, the designated SC9 secretariat, and Peter Greve, the chairman of CEN TC123 who would become a group leader in SC9, all attended the seminar. It was a valuable opportunity for the LEOMA leadership to interact with some of the important players in the new ISO Laser Committee.

The inaugural SC9 meeting took place during March 6–9, 1990, at a hotel adjacent to the San Francisco International Airport. Five engineers from France, 6 from Germany, 3 from Japan, 1 from the United Kingdom, and 28 from the United States participated. During the meeting, these engineers divided themselves into seven "Work Groups," each addressing a different area:

Work Group 1: Terminology, test methods, and test instruments

Work Group 2: Interfaces and system specifications

Work Group 3: Safety

Work Group 4: Lasers for medical applications

Work Group 5: Lasers for nonmedical applications

Work Group 6: Optical components for lasers

Work Group 7: Electro-optical systems other than lasers

The charter of Work Group 3 was tricky, because responsibility for laser eye safety had been delegated to TC76 at the Geneva meeting in December. But the European Union's Machine Directive mandated safety standards for

---

[4]LEOMA's executive seminars evolved into an important annual event, as described in Chapter 7.

machines that used lasers, specifically laser machine tools used to drill, cut, weld, and otherwise process materials. In Europe, the CEN Laser Committee—TC123—had taken on the task of creating that standard. Unlike the other standards being created in TC123, the machine-tool standard would be legally binding because it was mandated by the Machine Directive. The content of this standard was crucially important to U.S. manufacturers, because they would be legally blocked from selling their machine tools in Europe if the tools did not comply with the standard.

But because the standard was being drafted in TC123, the only way the United States could influence it was to draft a similar standard in SC9. Thus, Work Group 3 was created specifically to address that issue.

By the time the inaugural meeting was concluded, SC9 had begun work on 43 new draft standards. And the committee had agreed to a very ambitious schedule for developing those drafts into full-fledged international standards. Each draft begun at the meeting was to be crafted into a so-called working document by July. In ISO parlance, a working document is close to the final form of the standard, subject only to fine-tuning within the work group that created it. A second SC9 meeting was scheduled in Germany for October, and by February 1991, all the SC9 projects were to become Draft International Standards (DIS).[5] These Draft International Standards would then be forwarded to CEN TC123 where, the plan was, they would be adopted verbatim as CEN European standards.

Overall, the inaugural meeting of SC9 had been quite satisfying. Later in March, at a meeting of LEOMA's executive committee, I recommended that our arrangement with Clifford Chance, the Brussels attorney, be terminated. The committee accepted that recommendation, although Bob Pressley carefully noted in the minutes that hiring the attorney had been the correct decision in December, when it had been made. But subsequent events—my meeting Greve and Giesen in February, the interaction between the Europeans and LEOMA members at the executive seminar, and most importantly the successful SC9 meeting—had all given LEOMA confidence that we were getting a handle on international standards.

The handle was not without flaws, one of which became apparent during discussions between SC9 and TC123 in the late spring of 1990. While CEN was willing to accept most of the standards drafted by ISO engineers from many nations, they were not confident that those same engineers could create a safety standard that would meet the requirements mandated by the European Machine Directive. But U.S. manufacturers were adamant about having a role in creating a standard that would be crucial to selling their products in Europe.

---

[5]Again in ISO parlance, a DIS is extremely close to the final form of a standard, subject only to trivial modifications like punctuation, numbering, and so forth.

Jim Smith of IBM, a senior member of IEC TC76 who was also active in SC9, arranged a meeting in Boulder in June to seek a resolution of this troubling issue. He invited Herman Boch, who was leading the effort in CEN TC123 to create the safety standard, and Tom Lieb, the LEOMA consultant who was helping craft the SC9 version of the same standard. A dozen or so other experts, from Europe and the United States, also joined the discussion.

The meeting lasted two days and was quite successful. Although they started out far apart in the philosophy of the safety standard, by the end of the two days Lieb and Boch agreed that it was feasible to combine the two philosophies in a single document. They agreed to collaborate via fax during the summer, and meet again in late summer. That meeting was equally successful, producing a final draft that was acceptable to both committees. That document would join all the other standards SC9 was creating, and become a Draft International Standard by February 1991.

The LEOMA board, satisfied with the results of the association's initial foray into the world of international standards, began seeking a mechanism to lock in its gains. One possibility was to replace Sid Braginsky, who wanted to step down as the SC9 chair, with an individual more familiar with the concerns of U.S. manufacturers, and with laser technology in general. Braginsky was an able administrator, but he was not even vaguely familiar with lasers, and could not contribute to the technical discussions at SC9 meetings, or at joint meetings between SC9 engineers and experts from other standards committees.

Of course, the LEOMA board thought I was a reasonable candidate for the job, but the politics within ISO are complex. ISO committee chairs are not elected by the committees, but are appointed by the secretariat of the next committee up the chain of command. In the case of SC9, the next committee up the chain of command was the ISO Optics Committee (ISO TC172). The secretariat of that committee, Hilmar Ganser, was the leading ISO negotiator in the squabble between ISO and IEC over laser standardization. The December 1989 meeting in Geneva had resolved some of those issues, but there were still many others under debate.

The issue in Ganser's mind was the ambiguity of LEOMA's—and my— perspective on this squabble. Although LEOMA had argued at the Geneva meeting in favor of ISO's creating standards for basic laser parameters like power and beam propagation, we had initially advocated that all laser standardization take place in IEC. Moreover, LEOMA was continuing to financially support Jerry Glen as secretariat of TC76. Sid Braginsky, on the other hand, had a long history with ISO and no entanglements at all with IEC. Ganser requested Braginsky to retain the chair of SC9 at least until the issues between ISO and IEC were settled.

If having a foot in both camps made me unsuitable as an SC9 chair, it also positioned me to influence the ongoing negotiations between the two. I drew up a complex diagram with one column for ISO and the other for IEC. I tried to identify every task, subtask and sub-subtask that should be involved in laser standardization, and put each of them on one column or the other. This document went first to one side for modification, then to the other to have those modifications modified, then back to the first, and so on. In short order, it became known as the "laser demarcation document."

One of the areas most in dispute was the issue of laser safety, which had been left unresolved at the Geneva meeting the previous December. By autumn of 1990, SC9 and TC123 were cooperatively crafting a standard for the safety of laser machine tools, but many in IEC felt this project was a brazen intrusion into IEC turf. Emotions could be intense. In November, Ganser went out of his way to warn me against allowing IEC engineers to "infiltrate" SC9 meetings and "block or forestall the work of SC9." To the best of my knowledge, nobody at IEC ever contemplated such a move.

Nonetheless, the ongoing dispute was absorbing resources on both sides, and in an attempt to end it, the upper management of both ISO and IEC invited all participants to a second meeting in Geneva in December 1990.

The "demarcation document," by now a full page of 10-point type with one column for ISO and another for IEC, was the focus of this second Geneva meeting. It had been debated and fine-tuned on both sides of the Atlantic for months, but IEC personnel nonetheless introduced an entire revision of the document in Geneva. After an hour's heated discussion, both ISO and IEC agreed to sign the original version of the document. With regard to safety, that document specified that the IEC be responsible for "Standardization of safety aspects *except* lasers used in an industrial materials-processing environment," which would be the responsibility of ISO.

So, after nearly two years of discussion, the division of responsibility for laser standardization was finally settled.

Meanwhile, in October 1990—two months before the Geneva meeting— SC9 had held its second meeting in Heidelberg, Germany. Unlike the first meeting in San Francisco, the Heidelberg meeting was officially a joint meeting of SC9 and TC123. The European CEN engineers and scientists met together with the ISO engineers and scientists from many countries. These experts agreed to abandon some of the 43 projects begun in San Francisco, and to work together to advance others. But, in the most significant outcome of the Heidelberg meeting, CEN TC123 agreed to cease all independent work on its standards. They would abandon their original versions of those standards, and adopt verbatim the standards developed in SC9 as CEN European standards.

LEOMA had achieved its objective of protecting U.S. manufacturers from restrictive laser standards in Europe after 1992. At a meeting in November 1990, the LEOMA board unanimously agreed to extend the members' voluntary standards assessment into 1991, its second year.

Although work continued at an intense pace during the remaining months of 1990 and early 1991, only five of the SC9 projects had advanced to become Draft International Standards by the target date of February 1991. These five dealt with the following:

- Requirements for documentations (instruction manuals, etc.)
- Mechanical interfaces with lasers
- Damage testing of optical surfaces
- Terminology, symbols, and units of measure
- Optical components of lasers

Nonetheless, progress on these and other projects was sufficient to give the Europeans confidence that the CEN standards could be ready by the deadline at the end of 1992.

The third meeting of SC9 was held in Las Vegas in March 1991. For the first time, the Chinese participated along with their colleagues from the United States, Japan, and Europe. The committee was now actively working on 15 standards, in addition to the five that had already been completed. And, because the tension between ISO and IEC had been resolved, I took over as the committee chairman. Sid Braginsky, the previous chairman, moved up to become the chair of the ISO Optics Committee (ISO TC172).

As things continued to move smoothly through 1991, the question of the standards assessment came up at LEOMA board meetings. The assessment was raising well over $100,000 a year, but it was an expense companies were anxious to reduce now that international standards seemed less threatening. The problem was that LEOMA's income from normal dues was insufficient to support both me as chairman of SC9, and Jerry Glen as secretariat of TC76.

One possible solution was federal funding for the standards effort. My experience in SC9 was that participants from every other nation had some financial support from their respective governments. Only the U.S. delegation was supported completely by industry. A likely source of federal funds was the National Institute of Standards and Technology (NIST), a part of the Commerce Department. At Commerce, we knew Nancy Mason, who was the deputy undersecretary for technology and who had been a speaker at the LEOMA executive seminar[6] in March. In July 1991, Bob Pressley and I visited her in Washington to discuss the possibility of NIST's helping to fund LEOMA's activities in international standards. At her suggestion, I prepared a formal proposal for such funding. It was unsuccessful. In a formal response, Mason wrote that there had been congressional hearings into the possibility of

---

[6]The series of executive seminars is described in Chapter 7.

NIST support for U.S. participation in international standards, but those hearings concluded that federal financial support would undermine the concept of voluntary standards.

With a degree of reluctance, the LEOMA board in September authorized the standards assessment to continue into 1992, its third year. But the board emphasized its desire to find alternative funding mechanisms.

Meanwhile, at its fourth meeting, in Kobe, Japan, in November 1991, SC9 approved six more Draft International Standards, including one prescribing techniques to measure a laser's output power, energy, and the temporal characteristics of a pulsed laser. Significantly, also included among the six was the standard for safety of machine tools using lasers, the project that the Europeans had been reluctant to assign to ISO. The new ISO Draft International Standard met all the requirements of the European Machine Directive.

As we had agreed at one of the earliest organizational meetings, in February 1990, SC9 had been meeting every six months during 1990 and 1991. In Kobe, the Europeans acknowledged that SC9's progress would meet the deadlines set for the European Community by the end of 1992. We agreed to cut the frequency of the committee's meetings back to one a year.

Funding was still an issue. Those LEOMA members who voluntarily supported the standards assessment felt that, since the entire industry benefited from LEOMA's participation in the development of laser standards in both IEC and ISO, support for that participation should come from the broader industry and not just from a dozen LEOMA companies. Perhaps the problem was that other companies knew too little about LEOMA's success with international standards. I was directed to organize a seminar at the industry-wide OPTCON[7] conference in November, advertising LEOMA's progress to the entire photonics community.

The seminar featured nearly a dozen of the U.S. project leaders in SC9, each of them describing part of the overall SC9 mission. Attendance was sparse or, as Tom Galantowicz put it wryly in the minutes of the LEOMA board meeting several days later, "The lack of overflow would indicate that not many companies are aware of the critical nature of . . . international standards." It was one more manifestation of LEOMA's perennial problem: Companies saw no benefit in addressing an issue themselves when the LEOMA companies were already solving it.

We were indeed solving the problem of laser standards in both IEC and ISO, yet another standards issue came to the fore in early 1992. ISO 9000, the quality-management standard, posed another threat to sales throughout the world. Although the larger LEOMA members were already aware of the standard and its potential consequences, a survey of the membership in

---

[7]LEOMA's role in the OPTCON conference is discussed in Chapter 2.

general indicated the awareness had not spread to the smaller members. I was instructed to become knowledgeable about the subject, and to keep the membership informed via the LEOMA newsletter. In particular, I compiled a list of qualified ISO 9000 consultants the members might use, which was published in the newsletter.

In July 1992, LEOMA organized a members-only seminar about ISO 9000 at Spectra-Physics. The speaker was Robert Peach, who had chaired the ISO committee that had created the official "how-to" guide for ISO 9000. The seminar was well attended, and by the end of summer, a survey of the LEOMA membership indicated a much broader awareness of the standard: Fully 96% of the membership now planned to implement some form of ISO 9000.

As a part of its support of the 1992 OPTCON conference, LEOMA organized another ISO 9000 seminar at OPTCON in November. This seminar featured two different ISO 9000 experts, and was open to LEOMA members and nonmembers alike, at no charge.

But by late 1992, it was becoming obvious that ISO 9000 was not as appropriate an activity for LEOMA as the ISO laser standards. Companies that were well on their way to implementing the standard regarded their expertise as a competitive advantage, and were not anxious to share it with competitors just starting with ISO 9000. Moreover, the companies that were supporting LEOMA's work with laser standards—work that benefited the entire industry—were reluctant to once again shoulder the burden for the whole industry. The OPTCON seminar was the last significant activity LEOMA undertook in connection with ISO 9000.

By now on a schedule of one meeting per year, SC9 met in Paris in September 1992. The committee was finishing work on the last standards required by the Europeans for their end-of-1992 deadline, and began addressing new standards in areas like interferometry, optical-damage measurement, integrated optics, and safety of instruments and materials used in conjunction with laser surgery. By this time, employees of at least 15 U.S. laser companies[8] were involved in creating these standards.

Still unresolved was the issue of financing LEOMA's ongoing work in international laser standards. The LEOMA board was determined not to continue the expensive standards assessment into a fourth year. Feeling that the burden should legitimately be borne by the entire laser/photonics industry, the board decided to request that the two major laser conferences in the United States—CLEO and OPTCON—add a standards fee to every exhibitor at those conferences.

---

[8]Amoco Laser, Candela, Coherent, Hughes, IBM, Laser Photonics, Laserscope, Lightwave Electronics, Lumonics, Melles Griot, Photon, Quantronix, Rofin Sinar, Spectra-Physics, and Uniphase.

The requested fee was to be $60 for each 10' × 10' booth at both conferences, and LEOMA board members embarked on a concentrated lobbying campaign with individual members of the steering committees of the conferences. We enlisted the other trade association interested in optics standards, the American Precision Optics Manufacturers' Association (APOMA), to join us in making the request.

Because I was a member of the OPTCON steering committee, I formally presented LEOMA's request to the committee at the 1992 OPTCON meeting in November. I was joined by two LEOMA board members (Tom Galantowicz and Don Scifres) and two APOMA board members (Dick Sellers and Harvey Pollicove). The request was immediately opposed by the marketing people at OSA and SPIE, who argued that the additional fee would discourage exhibitors. After discussion, the steering committee tabled the request until its next meeting.

I also made the presentation to the CLEO steering committee during the November OPTCON meeting. The immediate reaction of the individual members was clearly negative. Tom Galantowicz, who was a LEOMA board member, was the only member of the steering committee to speak out in support of the proposal (see Figure 3.1). I left the meeting following my presentation, but was subsequently informed by letter that the steering committee had rejected the proposal by a near-unanimous vote. The letter from the steering committee chair, Jack Bufton, explained that the committee felt that standards were important to the entire CLEO community, but "CLEO was the wrong body to assess a tax on the industry."

At its meeting in January 1993, the OPTCON steering committee likewise rejected the LEOMA proposal. As reflected in Chapter 2, the level of dissention within the OPTCON steering committee was already pretty high, and the opposition to the proposal was vociferous and based primarily on the concern that a fee would have negative impact on booth sales at the OPTCON exhibition. In the end the vote was a tie, so the proposal failed because it lacked a majority.

So in the winter of 1992–1993, the LEOMA board faced the decision between abandoning its standards work or implementing a large increase in members' dues. They found a middle road. As distasteful as any dues increase was, the board approved an increase that was significant, but was not sufficient to replace the funds lost when the standards assessment ended. For LEOMA's smallest members, the boost was negligible: from $325 a year to $375. For the largest members, the increase was more substantial, from $14,950 to $25,000.

This dues increase, plus LEOMA's existing funds, was enough to keep the standards project going through 1993, but not indefinitely. One possibility for supplying the missing funds surfaced later in 1993, when LEOMA joined with many of the professional societies in a multiyear proposal to the Technology

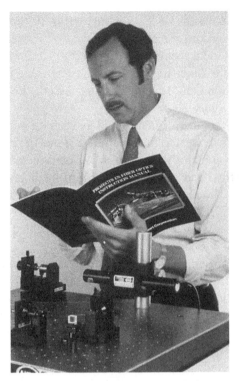

**FIGURE 3.1**   Tom Galantowicz strongly advocated OPTCON's and CLEO's support of international laser and optics standards.

Reinvestment Program at DARPA. The proposal requested several hundred thousand dollars to support a significant project in optics standardization in the United States. The main thrust of the proposed project was to convert existing optical military specifications ("mil specs") into optics standards that would be published by the American National Standards Institute, but the proposal also requested some $50,000 a year to underwrite SC9 activities.

At the time, there was considerable optimism that the proposal would be successful, but it was not. In the end, the funds lost when the standards assessment was discontinued were replaced by revenue from the short course, *Understanding Laser Technology*, as described in Chapters 1 and 4.

But the larger project—converting U.S. mil specs into up-to-date optics standards—remained unfunded and unaddressed. Funding of optics standards in the United States is an ongoing problem because neither business nor academia perceives much real need for optics standards. Existing, ad hoc measurement and specification techniques, supplemented in some cases by mil specs, were widely accepted in the United States, and formal standardization was—and often still is—seen as an unnecessary bother. LEOMA and its members initially became very engaged in laser standards when it appeared that European laser standards would limit access to the European market. But once LEOMA had eliminated that threat, the companies lost interest quickly. The U.S. contingent at SC9 meetings shrank from dozens at the initial meetings, to a handful by 1993 and 1994. The LEOMA board maintained a minimal presence in SC9 not because SC9 was perceived to be creating useful standards, but to ensure that SC9 did not create nationalistic standards that would impede international sales of U.S. manufacturers.

Nonetheless, some individuals worried that the optics mil specs, largely based on WWII technology, were becoming outdated and surpassed by other optics standards elsewhere in the world. In late 1994, the Optics and Electro-Optics Standards Council (OEOSC[9]) was created to energize the U.S. optics community to participate in all ISO TC172 standards, not just those created in SC9. Other TC172 subcommittees addressed topics ranging from ophthalmic optics to telescopes and microscopes and geodetic instruments to basic optical materials and components.

In particular, TC172 had created ISO 10110 describing many basic techniques of optics drawings and specifications, a standard much more appropriate for the twentieth and twenty-first centuries than the ancient mil specs. And ISO 10110 had been created without input from the United States. "10110 is neat," Harvey Pollicove, then head of the Center for Optics Manufacturing told me at the time, "But it would have been a lot neater if the U.S. had played a role in creating it."

And the National Ignition Facility, a major program to develop laser-driven nuclear fusion at the Lawrence Livermore National Laboratory, had announced it would buy many millions of dollars worth of optics specified according to ISO 10110.

Despite the importance of ISO 10110 and other TC172 standards, OEOSC struggled and nearly ceased to exist in its early years. If need be, LEOMA could have supported U.S. activity in SC9 by itself, but there would have been no U.S. involvement in any of the other TC172 subcommittees. Only after Pollicove and others broadcast the direst warnings about consequences of this eventuality did the funding materialize to see OEOSC through to its second birthday. Even today, it survives year-to-year with minimal support from industry and academia. Standardization is simply not a priority to most of the U.S. optics community.

Another issue with European standards, completely unrelated to the work in TC172, came to the fore in 1995. The Electromagnetic Compatibility (EMC) Directive required manufacturers to meet its specifications in Europe by the end of the year. LEOMA took a survey of its members, and the results were

---

[9]The acronym reminds some people of Old MacDonald and his farm.

appalling; many LEOMA CEOs were unaware of the EMC Directive, and many of those who had heard of it were unaware of its consequences. A typical remark on the returned survey was, "That's something the power-supply manufacturers have to worry about."

The EMC Directive was something that any manufacturer of electronics equipment sold in Europe had to worry about. Failure to comply would bar products from the European market. After analyzing the survey, LEOMA launched a full-court press to ensure that its members were in compliance with the EMC Directive by the deadline of December 31, 1995. LEOMA organized and widely advertised a seminar on the subject at the 1995 CLEO, with speakers from one of the major compliance testers, TUV Rhineland, explaining the EMC Directive's requirements. LEOMA opened the seminar to any company that wanted to attend, but the seminar fee was significantly greater for nonmembers ($225 vs. $600). To the best of my knowledge, by the time December 31 rolled around, all LEOMA members were in compliance.

The year after the European EMC Directive went into effect, another EMC-related issue came up in SC9. I think sometimes standards writers try to standardize too much, and perhaps this is a case in point. The delegates from one of the SC9-member nations came to the 1996 SC9 meeting armed with a firm proposal for standardizing electromagnetic compatibility of lasers. The proposal outlined the amounts of radio-frequency radiation lasers would be allowed to generate, and the amounts they should be able to withstand without malfunctioning. Of course, these were the same quantities standardized by the European EMC Directive. In some cases the SC9 proposal duplicated the EMC's criteria, in other cases the SC9 proposal's requirements were stricter.

In my mind, the proposal was completely unnecessary and counter-productive. The EMC Directive would apply to lasers anyway, and another EMC standard would be confusing and add both complexity and cost to the laser marketplace. Fortunately, many of the SC9 engineers from other countries agreed, and after an hour's enthusiastic debate, the nation that had originated the EMC proposal agreed to drop it.

But a more serious problem appeared at about the same time as the EMC proposal in SC9. We learned that another standards committee, the IEC Semiconductor Committee, was creating standards for measuring the optical characteristics of semiconductor lasers. These standards addressed many of the same characteristics addressed by existing SC9 standards, but often described different measurement techniques. This was a disturbing development, because SC9 standards were intended to apply to all lasers, including semiconductor lasers.

Because the IEC Semiconductor Committee (IEC TC47) was comprised almost entirely of Japanese engineers and scientists, our first reaction was to request that the Japanese representatives in SC9 dissuade their countrymen in TC47 from pursuing optical standards. That course proved to be ineffectual.

Later in 1996, I joined Akira Arimoto, the leader of the Japanese delegation to SC9, and Karlhanns Gindele, the SC9 secretariat, in drafting a formal statement of SC9's position regarding TC47's optical-measurement standards for diode lasers. We began by outlining the multiple reasons why such activity was inappropriate. First of all, it violated the 1990 agreement between ISO and IEC that unambiguously delegated creation of optical standards for *all* lasers to ISO TC172 SC9. But even aside from that agreement, having two sets of measurement standards for lasers would introduce ambiguity among scientists and engineers working with lasers, and between laser manufacturers and their customers. The ultimate result would be impeded development of semiconductor lasers and their applications.

Moreover, we pointed out that the engineers and scientists in TC47 were experts in semiconductor technology, but there was very limited optical expertise in that committee. Finally, because TC47 had almost no input from European countries or the United States, they could not create optical standards that would achieve global acceptance.

Then, the document proposed what we considered to be a very positive resolution to the conflict. We acknowledged that SC9 had limited electronics and semiconductor expertise, so SC9 would gladly accept TC47's standards for measurement of electrical characteristics of semiconductor lasers. Conversely, we asked that TC47 accept SC9's standards for the measurement of optical characteristics. Those few characteristics that could be considered both optical and electrical—laser lifetime, for example—would be addressed by a special joint committee with experts from both TC47 and SC9.

To us, the document set forth a very strong case that would be difficult for any logical person to dispute. We forwarded the document to the TC47 leadership, and also to the ISO and IEC central offices in Geneva.

In reality, the document was largely ignored. In the spring of 1997, administrators in the IEC central office in Geneva issued a unilateral proclamation that IEC was authorized to create all standards pertaining to semiconductor lasers. From SC9's perspective, this proclamation was an act of extreme arrogance. I made a formal request to the ISO central office in Geneva to reconvene the Laser Coordination Group, the committee that had generated the 1990 agreement between ISO and IEC governing creation of laser standards.

There was reluctance of the ISO central office to become involved in squabbles between committees, especially since the IEC central office had already staked out a position in the matter. They suggested instead a face-to-face meeting between TC47 and SC9 to resolve the issue. Accordingly, I invited TC47 to meet in conjunction with the SC9 meeting in Maurach, Germany, in June 1997.

That meeting was attended by the TC47 secretariat and the administrator who handled TC47 affairs in the IEC Geneva office. It was not a successful meeting. We essentially reiterated the same arguments that we had made in the formal letter Arimoto, Gindele, and I had written a year earlier—the arguments we felt no reasonable person could dispute. The gentlemen from IEC didn't bother disputing them; they simply insisted that SC9 standards did not apply to semiconductor lasers, and therefore TC47 would create the necessary standards. I guess we talked at each other for several hours, without making any headway. After the meeting concluded, our German hosts took everybody to dinner, where we were uncomfortably polite to each other.

During the ensuing months, I continued to press the ISO central office to reconvene the Laser Coordinating Group, and continued to encounter reluctance. Nobody said "No," but the matter just kept slipping sideways. The United States is officially represented in ISO by the American National Standards Institute (ANSI) in New York, and I contacted the ANSI officials who participate in ISO issues and urged them to lobby for reconvening the Laser Coordinating Group. More sideways slipping.

The 1998 SC9 meeting was held in the village of Rusutsu, Japan, the homeland of most TC47 members. What happened there is described in the following paragraphs from the autumn 1998 LEOMA newsletter:

> As a prelude to [the SC9 meeting in Rusutsu], members of the ISO committee sat down with members of the IEC committee to take another stab at resolving the dispute.
>
> And there was no dispute.
>
> Members of the two committees readily agreed that the ISO standards would be expanded to include measurement techniques of concern to the IEC, and that the IEC committee would not duplicate or contradict any standards already put in place by the ISO.

In other words, 30 months of dissention had been fueled not by the technical experts doing the real standardization work, but by the administrative personnel who were supposed to support the experts. Once the ISO engineers and scientists sat down with their IEC colleagues, all problems dissolved. It was a remarkable revelation.

After the Rusutsu meeting, SC9 set about the tasks it had agreed to in Japan, expanding the ISO standards to accommodate measurement techniques appropriate for semiconductor lasers (e.g., adding the possibility of using in integrating sphere to measure the output power of the highly divergent beam from a semiconductor laser). And I continued to press, with renewed enthusiasm now, for a meeting of the Laser Coordinating Group in order to put an official and final stamp of approval on the Rusutsu agreement.

That meeting was finally scheduled in Geneva for June 1999. But several weeks before, that meeting was inexplicably cancelled. And, as we later learned, an unannounced meeting between ISO and IEC administrators in Geneva replaced it. And at that meeting, the administrative staff agreed to recommend to the ISO Technical Management Board (ISO TMB)—the ultimate authority within ISO—that ISO be prohibited from creating standards for diode lasers.

The word "shocked" is not incorrect in describing SC9's reaction to this development. In my exasperation, I faxed an angry letter to the ISO Central secretariat, accusing him of ignoring the opinions of his own experts within ISO as well as those of the IEC experts. It was an impolitic letter and, in retrospect, I imagine it did more harm than good.

The other moves I made were wiser. I traveled to New York, where I met with the ANSI officials who represented the United States on the ISO TMB. I successfully urged them to vote against the recommendation that the administrators would place before the board. And I encouraged my SC9 colleagues from other nations to do the same with their respective national standards bodies.

All this lobbying apparently paid off, because the TMB roundly rejected the recommendation from the administrative staff. In the years that followed, the Rusutsu agreement guided the interaction between the IEC Semiconductor Committee and the ISO Laser Committee. Even today (2014), a joint working group comprising technical experts from both committees meets occasionally and cooperatively to address issues where the committees' charters intersect.

Since its inception in 1990, the ISO Laser Committee has created more than three dozen international standards that apply to lasers and electro-optical systems. The broad topics covered by these standards are listed in Appendix 2. Although LEOMA ceased operations in 2004, I continue to chair SC9, and the expenses are covered by an informal arrangement with two major laser companies, Newport and Coherent.


## THE BENEFIT OF HINDSIGHT

LEOMA was fortunate to find so quickly the path to successful interaction with the European standards organizations. Many paths were available at the beginning, and more than anything, it was the helpful cooperation of the Europeans themselves who took us in the right direction. For more than

20 years now, the U.S. laser industry has been effective by placing experts directly in the ISO and IEC committees, instead of first creating domestic standards and then working to have those standards adopted by international bodies.

The pettiness we discovered in some corners of the standardization bureaucracy was disappointing. In LEOMA's case, we were able to sidestep some of the turf wars by detouring around the administrative staffs and going directly to the scientists and engineers who create the standards in other committees. But I imagine there are other cases where the scientists and engineers are, themselves, the guardians of fiefdoms.

LEOMA has always approached standardization from a defensive perspective. The reality is, nobody on the LEOMA board had much use for standards. Standards were viewed by the LEOMA executives as technically unnecessary—the existing measurement techniques were considered perfectly adequate—and administratively expensive.

As the Europeans were forging their new standards for "Europe '92," the assumption was that customers in Europe and elsewhere would demand that lasers be specified according to these new standards. For the most part, that has not been the case. Today, U.S. laser manufacturers rarely use the ISO laser standards in describing their products' performance, and nobody seems to mind. Manufacturers hedge their bets by maintaining a minimal participation in the various laser-standards committees, but that participation is purely defensive in nature.

# 4

# EDUCATIONAL ISSUES

From the very beginning, it was obvious to the LEOMA Board and its members that education, and especially science education, should be a priority for the association. As early as March 1989, the LEOMA board was evaluating specific projects that would boost the quality of education in the United States. One possibility was with the physics department at San José State University, where Professor Gareth Williams ran summer workshops for local high school science teachers. The workshops presented teachers with a series of experiments, many dealing with the principles of optics that the teachers might use in their own classes. Dale Crane, LEOMA's 1990 president, visited Williams and suggested several possibilities, including the possibility that LEOMA companies might donate equipment and/or employees' time to augment the summer workshops.

But Williams had a greater need, and what eventually came to fruition was a different project. Williams had recently launched a program he called LASE, in which local high school students were given a goal that might be achieved by utilizing a laser and other equipment the university provided. But the design of an experiment was up to the student. LEOMA wound up underwriting a publication, *LASE LOG*, which was distributed throughout the San Francisco Bay area and which described the students' successful experiments. My understanding is that *LASE LOG* was, for several years, more prestigious than *Physical Review Letters* among the techies in Bay area high schools.

*LEOMA and the U.S. Laser Industry: The Good and Bad Moves for Trade Associations in Emerging High-Tech Industries*, First Edition. C. Breck Hitz.
© 2015 by The Institute of Electrical and Electronics Engineers, Inc. Published 2015 by John Wiley & Sons, Inc.

LEOMA also cooperated with the professional societies in many of the educational activities they organized. In 1992, for example, both SPIE and OSA requested LEOMA to cosponsor separate "Education Days" they were presenting at that year's OPTCON.[1] LEOMA agreed to cosponsor both projects.

In 1994, LEOMA adopted my course, *Understanding Laser Technology*, and sponsored numerous public and private presentations during the ensuing years. The 18-hour course, which I had developed and taught many times before joining LEOMA, began with an explanation of atoms and electrons and atomic energy levels, but later explained—intuitively, without any mathematics—more sophisticated topics like the difference between homogeneous and inhomogeneous broadening of laser lines, and quasi-phase matching in nonlinear optics. Altogether, LEOMA, with an assist from Laurin Publications, sponsored presentations of the course to hundreds of technicians, engineers, and businesspeople around the United States.

Throughout the 1990s there was an increasing awareness that American science and math education in general, and optics-related education in particular, was failing to provide the skilled engineers, scientists, and technicians that the country would need in coming decades. In 1999, a LEOMA committee that had been formed to find ways the association might address the issue made a strategic proposal: LEOMA would produce a sophisticated PowerPoint presentation, suitable for high school students and college freshmen, describing the many career opportunities and optics and photonics. Each LEOMA board member would make the presentation at least twice a year at local schools. Moreover, the PowerPoint file would be widely distributed to the professional societies and their members, so the presentation could be made many dozens of times a year. The LEOMA Board embraced the project, and the 1999 president-elect, Bob Phillippy, agreed to spearhead the undertaking.

Although the concept had been for the project to be a collaboration among several LEOMA companies, Phillippy shouldered most of the burden himself. He researched employment data, acquired videos and other material, and brought everything together into a single PowerPoint presentation (Figure 4.1).

Technical problems—getting the presentation and its embedded videos to run smoothly on a variety of computers—delayed wide distribution of the presentation until early 2001. But by 2001 the economic situation had deteriorated significantly from previous years. The explosive growth the industry had experienced in the late 1990s had diminished drastically. Instead of being anxious to hire new workers, some optics companies were letting

---

[1] OSA and SPIE are two professional societies, and OPTCON was a large annual technology meeting. OPTCON and the professional societies are discussed in Chapter 2.

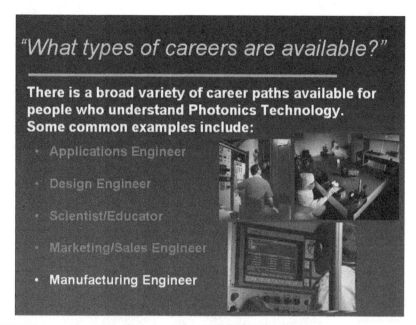

**FIGURE 4.1**   The LEOMA presentation for high school and college students described the exploding demand for engineers and technicians trained in photonics.

employees go. Although Phillippy and half a dozen others used the Power-Point file at local schools, the presentation never reached the enormous audience originally envisioned (see Figure 4.2).

But the PowerPoint presentation was not LEOMA's only response to industry growth in the late 1990s. The worker shortage was especially acute in the San Francisco Bay area, home to many of LEOMA's key members. Area laser companies projected a need for more than 300 new, trained laser technicians by 2003, and local schools were producing only a handful.

But the most pressing need was among the companies manufacturing fiber-optic components. The fiber-optics market was growing exponentially as telecom companies installed fiber-optic cables between cities and underneath city streets throughout the country and around the world. And fiber-optic vendors were desperate for employees. The Optical Fiber Conference, which had been a small academic conference until the late 1990s, was suddenly bursting its seams. I distinctly recall an OFC meeting in the late 1990s, in Baltimore, Maryland, that was probably the most overcrowded technical conference I have ever attended. And though the technical sessions were jammed, the most telling comments were those I heard between the sessions: "Everybody here's talking about the technology, nobody's talking about the real problem: *people*." "We're growing as fast as we can hire people." "I have to pay through the nose for optical engineers."

**FIGURE 4.2** Bob Phillippy, LEOMA's 2000 president and CEO of Newport Corp., almost single-handedly created the PowerPoint slideshow encouraging high school seniors and college freshmen to consider a career in optics and photonics.

## YUBA COLLEGE

Two of LEOMA's largest members, Coherent and Spectra-Physics, had established fiber-optics plants in California's Central Valley—in Auburn and Oroville, respectively—and were finding their growth seriously stunted by a lack of skilled employees. Exacerbating the problem, the nearby National Ignition Facility at the Lawrence Livermore National Laboratory projected a need for more than 200 optics technicians in the coming years.

If you connect Auburn, Oroville, and Livermore with dotted lines on a map of California, you will form a triangle. And within that triangle is Yuba College, a two-year community college in the town of Marysville. In early 1998, LEOMA contacted officials there inquiring about the possibility of creating a two-year program in optics fabrication, leading to an Associate of Science (AS) degree. Such a program, turning out dozens of graduates a year, would at least put a dent in the shortages.

But such a program would also be expensive. It would need a fully equipped optical-fabrication laboratory, where students could learn and practice the shaping, grinding, and polishing techniques necessary to fabricate precision optics. Such a laboratory could easily cost tens or hundreds of thousands of dollars, and the officials at Yuba initially rejected LEOMA's suggestion of an optics-fabrication curriculum at that school.

LEOMA responded by proposing that the U.S. optics industry, plus Livermore, would provide the equipment if Yuba provided classroom space and had the curriculum formally approved by the state of California. That proposal led to the first serious meeting among Yuba officials and representatives of LEOMA, Coherent, Spectra-Physics, and Lawrence Livermore in July 1998.

The project moved quickly after that first meeting. By autumn of 1998 a curriculum had been outlined, the necessary laboratory equipment defined, and potential instructors identified. The plan was to launch the program with a full incoming class of 20 students in August 1999. "This program is evolving much faster than I dared hope," enthused George Balogh, head of Spectra-Physics' optics division.

Momentum continued to build. In early 1999, Vicki Hoffman, Coherent's representative to the Yuba committee, delivered a letter signed by Bob Gelber, head of Coherent's optics division, pledging up to $100,000 to support the Yuba program. Spectra's Balogh immediately matched the pledge. With $200,000 in cash support pledged from industry, plus promises of wide-ranging equipment donations, plus the potential of state funding being pursued by Yuba, the optics program was steamrolling ahead.

Of course, there would be bumps in the road. Finding suitable laboratory space was one of the first. It turned out there was absolutely no space anywhere on the Yuba campus. Emergency classroom space can often by supplied with temporary buildings, but that would not work for an optics lab because the floor of a temporary building cannot support the massive equipment needed to fabricate precision optics.

For a time, Coherent and Spectra considered constructing a building on the Yuba campus at their own expense. Initial concepts of the lab called for 500 sq. ft, and local building costs were running at $50–70 a sq. ft, so the expense of $25,000–35,000 seemed to be within reason. There was discussion about whether the companies should build and own the building, or should simply route the funds to Yuba, so Yuba could build the building.

But as the role of the laboratory was refined, the required size grew first to 800 sq. ft, and then to 1500 sq. ft. And because of existing union contracts, the cost of construction on campus grounds was significantly higher than the prevailing cost elsewhere in the region. The concept of building a new building was proving prohibitively expensive.

Mike Moyers, Yuba's dean of Vocational Instruction, came up with a solution. The Feather River Shopping Center, a few miles from the campus, had failed as a shopping center and was being renovated as office and classroom space. Several of Yuba's programs were moving to that facility, and space for the optics lab was available for approximately $15,000 a year for the first two years, after which Yuba College would be able to fund the costs from tuition income.

The curriculum was another issue. Although there are numerous curricula for fundamental optics courses, what was needed at Yuba was instruction in the fine art of fabricating precision optics. The requisite expertise certainly existed at both Coherent and Spectra-Physics, but distilling that expertise into a teachable curriculum was a daunting task.

Spectra-Physics nominated one of its employees, Gordon Soekland, to be the instructor. Soekland had been a master optician at another company, Optical Coating Laboratory Inc. (OCLI) for 22 years before moving to Spectra-Physics' optics division. Decades earlier, when Soekland had started at OCLI, he had taken an optics-fabrication course developed by a pair of OCLI experts, Gerhard Wolf and Hank Karow. Starting with the rudimentary outline of the curriculum of that course, Soekland and Yuba's Mike Moyers developed a one-year curriculum suitable for Yuba College.

With classroom and laboratory space located and a curriculum identified, the next problem was equipping the laboratory. Soekland put together a wish list of the machines he needed to teach the course, and Spectra, Coherent, and Livermore found that they could supply much of it, mainly older models that were still suitable for teaching. Yuba's financial offices set out on a quest for state funds to help with the equipment.

The equipment and supplies that weren't donated had to be purchased. Coherent and Spectra agreed to deposit funds—initially $15,000 apiece—in a LEOMA account. Soekland would get approval from both companies to purchase specified equipment, and the invoices would go to LEOMA.

But there was a setback when Bob Gelber, who had headed Coherent's optics division, retired and John Ambroseo took control of the division. After a visit to the Yuba campus in 2000, Ambroseo would become an ardent supporter of the optics program. But in May 1999, he was skeptical about the program in general, and in particular he questioned whether it should cost $200,000 to equip a lab where basic optical-fabrication techniques were taught. He reduced the $100,000 pledge that Gelber had made to $25,000. George Balogh likewise cut Spectra's pledge to $25,000.

There was consternation and regrouping as the news was absorbed at Yuba. It was only a little more than two months until classes were scheduled to start. Soekland revised his plan by substituting Costco tables and Kmart hotplates for the industrial-grade equipment in the original plan. Still, after the plan was fully revised, the consensus was that the cobbled-together lab would suffice for the first year, and better equipment could be slowly integrated as the program matured.

The curriculum that Soekland and Moyers had put together earlier in the year was formally approved, both by Yuba College and the state of California, well before classes were scheduled to start.

The effort to bring students into the program began early in 1999. About a quarter of Yuba's students were recent high school graduates, and the remainder were adults returning to college for further training in hopes of raising their incomes. We assumed that the optics program would comprise a similar mix, and designed the outreach to that audience.

Moyers crafted a survey, which he circulated to 51 Bay area companies LEOMA had identified as likely to employ graduates from the program. The feedback indicted that well-paid jobs would be plentiful, and this result was featured prominently on brochures and flyers distributed to potential students. Career councilors from local high schools were invited to visit the Coherent and Spectra-Physics optics facilities, where they learned of the opportunities for trained opticians. At Yuba's Orientation Day in 1999, where incoming students learn about the college's different programs, human resources representatives from both companies were present to greet the new students and acquaint them with Yuba's optics program—and the opportunities awaiting them upon graduation.

Everything came together, miraculously, in August 1999, when 23 students—the full capacity—enrolled in the optics program. Another half dozen students were on the waiting list. There was more than enough equipment available to begin the laboratory portion of the curriculum. Coherent and Spectra-Physics together had donated some $67,000 worth of fabricating machinery, Livermore added another $90,000 worth, and other optics companies around the United States had donated some $40,000 in equipment. A plaque prepared by Yuba College for the classroom listed the donations to the program; Table 4.1 is a transcription of that plaque. But that equipment sat idle for several months, because the Feather River Center was still under construction and the optics laboratory could not open for business until well into the first semester.

**TABLE 4.1    A Plaque on the Wall of the Yuba College Classroom Listed the Many Contributors to the Optics-Fabrication Program**

CNC curve generator, Coherent Optics, Auburn, CA
Wyko Optics Profilometer, Coherent Optics, Auburn, CA
$30,000 for program initiation, Coherent Optics, Auburn, CA
Tooling for Optics Manufacturing, Coherent Optics, Auburn, CA
Wyko 400 Interferometer, Bosby & Assoc, Auburn, CA
Pitches, waxes, cleaning agents, Universal Photonics, Folsom, CA
Glass blanks for blocking plates, United Lens Co., Southbridge, MA
Lens design for telescope project, Wavelength Optics, Sacramento, CA
Lens centering machine, LaCroix Optics, Batesville, AZ
Strasbaugh polisher/grinder, Coherent Optics, Auburn, CA

**TABLE 4.1**   *(Continued)*

Breadboards for testing platforms, Coherent Optics, Auburn, CA
$30,000 for program initiation, Spectra-Physics, Oroville, CA
Tooling for optics manufacturing, Spectra-Physics, Oroville, CA
Polishing rods, slurries, supplies, Rodel Optics, Danville, CA
Glass blanks for projects, Schott Glass, Fullerton, CA
Measurement tooling, Avantech Machining, Auburn, CA
Glass substrates, Newport Glass, Costa Mesa, CA
Precision optical saw, LaCroix Optical, Batesville, AZ
Spherometer rings for testing, LaCroix Optics, Batesville, AZ
Cast tooling blanks, Optical Coating Labs, Santa Rosa, CA
Air isolation table, Lawrence Livermore National Laboratory, Livermore, CA
Blanchard #18 grinder, Lawrence Livermore National Laboratory, Livermore, CA
Consulting & advisory assistance, Coherent Optics, Auburn, CA
Consulting & advisory assistance, LEOMA, Pacifica, CA
Storage shelving, Optical Coating Lab, Santa Rosa, CA
Granite measuring tables, Lawrence Livermore National Laboratory, Livermore, CA
Air-flow bench, Lawrence Livermore National Laboratory, Livermore, CA
Precision optical glass, Lawrence Livermore National Laboratory, Livermore, CA
Consulting & advisory assistance, Spectra-Physics, Oroville, CA

The expectation was that students completing the first year—the optics-intense year–would be awarded a certificate for that achievement, and some of them would immediately begin work for optics companies. Others would stay in school for a second year, taking general science, math, and humanities courses, and be awarded the Associate of Science (AS) degree.

Potential employers included not just Spectra, Coherent, and Livermore. These sponsors of the Yuba program realized it was crucial that those completing the program have a range of options, and optics companies across the United States were invited to participate in recruiting at Yuba. "The goal is to create dynamic program . . . that attracts students by offering a wide range of employment opportunities," Coherent's John Ambroseo commented at the time (see Figure 4.3). Spectra's Greg Sanger echoed the sentiment: "I certainly hope to see a lot of companies recruiting at Yuba."

But as the program progressed, there were further difficulties. Despite Mike Moyers's pledge earlier in 1999 that Yuba would fund the course instructor, the funds did not materialize and Soekland wound up being paid mostly out of funds provided by LEOMA. As a result, the LEOMA-supplied pot of money was being depleted more rapidly than planned, and the companies began pressing Yuba to provide better support to the program.

Spectra's Greg Sanger, in particular, argued that Soekland should become a regular Yuba faculty member, paid from the funds Yuba collected from students. But there were "procedural/union difficulties" with that concept,

**FIGURE 4.3**    John Ambroseo at a CLEO conference in about 1998. Ambroseo was a strong supporter of the optics training curriculum at Yuba College.

Moyers explained. And Soekland's compensation continued to come from Coherent and Spectra-Physics.

Further support from Coherent and Spectra came in the form of paid internships for students enrolled in the Yuba program. (Of course, these interns also supplied badly needed labor for the companies.) The companies offered part-time internships to students during the school year, as well as full-time internships during vacation periods. In September 1999, two of the Yuba students immediately signed up for the internships, and by December there were half a dozen. Moreover, several of the students who had enrolled in September simply dropped out of the program and went to work full-time at the companies.

To replace those who had gone to work, several students from the wait list were admitted to the program at the beginning of the second semester. In the second semester, Soekland again taught his introductory optical-fabrication course, and another instructor, Hank Karow, taught optics fundamentals to the second-semester students who had taken Soekland's course in the first semester. Karow was one of the engineers who had developed the basic optics course at OCLI years earlier. And, despite the companies' displeasure, Karow's compensation also was paid from the LEOMA-provided funds.

The original intention had been to offer a second year in the curriculum, so students wishing to earn an AS degree could do so. But the optics companies were ravenous for trained opticians, and the starting salaries were correspondingly generous. By the spring of 2000, it became obvious that none of the students slated to receive certificates in June would turn down those enticing salaries and stay in school for an AS degree. There was no need to design a curriculum and identify instructors for the second year. Instead, we set about expanding the one-year program for the 2000–2001 school year.

Meanwhile, as the first year of the program approached its conclusion with almost 20 students expected to receive certificates, optics companies were invited to a recruiting night in May, where they could meet and make offers to the students. Fourteen companies attended that event, and in June all but two the graduates immediately began work. (One student had become pregnant, and one had a death in the family.)

Happily, Yuba had received a $250,000 grant from the state of California to support the program, so prospects for expanding the program were good. Part of that money went toward the purchase of six new spindles, so at the beginning of the second year, the optics lab was equipped with

- 21 spindles,
- 3 curve generators,
- 2 Zygo interferometers,
- 1 Meyer/Burger optical saw, and
- 1 Bridgeport horizontal mill.

Forty new students were enrolled at the beginning of the program's second year, in the autumn of 2000. And finally, after a year's delay, Gordon Soekland became a full-time member of the Yuba College faculty. Hank Karow continued teaching at Yuba on a part-time basis, and Norm Thomas, a Livermore employee, taught a course in geometrical optics that was televised from Lawrence Livermore National Laboratory to Yuba.

From the beginning, LEOMA's intention had been that optics fabrication would become a self-supporting course of study at Yuba College, like auto mechanics or bookkeeping. But for reasons that were never clear to me, that seemed to be a very difficult goal to accomplish. Even with a laboratory full of donated equipment and a $250,000 grant from the state, the school still needed extensive financial support from the LEOMA companies. Reluctantly, Spectra-Physics and Coherent pledged another $31,000 apiece for the 2000–2001 school year.

But by late 2000, disturbing hints that the wildfire growth of fiber optics were slowly beginning to circulate. The October 19, 2000, edition of the *Wall Street Journal* reported that many investors were shying away from companies that provided "bandwidth" with global networks of optical fibers. New technologies were enabling existing networks to handle an ever-increasing load. Some analysts, the report said, were predicting a capacity glut and falling prices.

The LEOMA companies, still desperate for new employees, worried that such reports could impede progress of the optical-fabrication program at Yuba. At Coherent and Spectra-Physics, orders were still coming in faster than the existing employees could fill them. At a meeting at Yuba College in early 2001, I cautioned against overreacting to such news. Sales data from companies that marketed fiber-optic networks remained strong: Nortel Networks, for example, a leading supplier of fiber-optic networks, had reported a 90% increase in sales from the third quarter of 1999 to the third quarter of 2000. *Laser Focus* magazine reported that market for telecommunications lasers had grown to over $5 billion by the end of 2000, a growth of 132% compared to the previous year.

Buoyed by optimism from Spectra and Coherent, the Yuba program gained strength. We discussed expanding it to other schools and, indeed, a similar program, described below, was already being organized at Irvine Valley College in southern California. Several Yuba students had dropped out of the program during the fall semester, but they were replaced by students from the waiting list, bringing the total enrollment up to 48.

At Yuba College, the new optics program was viewed as being very successful, even while other programs were losing enrollment. Further expansion of the optics program was limited by the lack of qualified instructors, and the engineering school suggested that several of its instructors might teach courses in the optics program. LEOMA and the companies discouraged this, however, saying that the optics courses required specialized knowledge and experience. It was true: all the optics instructors–Soekland, Karow, and Thomas–had extensive career experience with optics and optics fabrication.

In the most general terms, fabrication of precision optics is a two-step process. First, the part is cut, ground, and polished to the appropriate shape. Then a sophisticated thin-film coating is applied, determining the part's optical characteristics at specified wavelengths. Because the equipment to apply the thin-film coatings is prohibitively expensive, the Yuba program had addressed only the first step, but in the spring of 2001, plans were being laid to expand the program to include thin-film coatings. The classroom work would take place at Yuba, and the students would gain hands-on experience by serving internships at the optics companies.

But discouraging news from the outside world continued to filter into the Yuba program. Financial analysts continued to issue their bearish predictions on fiber-optic networks. And then a press release from JDSU, one of the leading suppliers of fiber-optics components, announced a layoff of optical technicians. Plans for initiating a thin-film class at Yuba in the fall of 2001 were put on hold.

During the spring of 2001, the slowdown became increasingly apparent. As orders slowed, Spectra and Coherent no longer needed new employees. And then, to everybody's dismay, some of the recent hires from the Yuba program were let go.

By the end of the spring semester, the picture was grim. Of the 48 students enrolled in the optics program at the beginning of the semester, nearly half had dropped out. Sixteen received certificates and sought employment, but only three had been hired. Worse, of the 18 students who had been hired at the end of the 1999–2000 school year, 9 had subsequently been laid off.

Was the downturn a mere bump in the road, or was the market for fiber-optic components collapsing? There were hopeful signs that it was merely a bump in the road. George Balogh pointed out in July that Spectra's optics division had grown from 135 employees at the end of 1999 to 400 at the end of 2000, and had subsequently shrunk to 230. The company fully expected its employment to grow to 400 and beyond in the near future, Balogh said.

Several people brought data from a recent fiber-optics conference organized by the marketing firm RHK Inc. to that July meeting at Yuba. According to RHK, Internet traffic was continuing to grow at 100% a year. A speaker at the conference, Paul Lacourture, president of Verizon Networks, had said that Verizon saw no decrease in the growth of demand for bandwidth. Overall, the conclusion from the RHK conference was that demand for bandwidth was growing steadily, but network companies had overbuilt in 2000, leading to a slowdown in 2001. But strong growth was expected to resume in 2002 as the oversupply from 2000 disappeared.

The Yuba program rallied around the hope that the downturn would be temporary. With the pressure to train workers for immediate employment abating, the focus turned to keeping students in school longer, so they could earn an AS degree.

Still, the politics at Yuba were becoming dicey. Soekland had been appointed a full-time member of the Yuba faculty, but he was being paid out of the $250,000 grant the program had received from the state, not from Yuba's general funds. Dean Moyers asked Spectra's and Coherent's representatives to attend a Yuba board meeting to reaffirm their companies' support for the program. That reaffirmation would inform the board's decision on whether Soekland could be paid from general funds after the grant expired at the end of the 2001–2002 school year.

That meeting did not go smoothly. Vicki Hoffman from Coherent and George Balogh from Spectra-Physics were not in a position to make ironclad guarantees of continuing the heavy financial support after 2002. They emphasized that the companies had already poured well over $100,000 cash into the program, in addition to tens of thousands of dollars worth of equipment and countless person-hours in planning and guiding the program's development. They wanted Yuba to provide more support. "It was our last-ditch effort to get Yuba to provide sufficient funds to the program," Balogh remembered in a recent conversation.

The Yuba Board of Trustees, on the other hand, felt the companies were not doing enough. The term "corporate greed" was inserted into the discussion, as the companies' motivation for participating in the program in the first place. That did not sit well with Balogh and Hoffman, and the meeting ended on a sour note. There was no resolution about Soekland's salary after 2002.

In an attempt to smooth over the political friction, Balogh, Hoffman, and I met for lunch with Yuba's president, Stephen Epler. There was a real possibility of the companies' withdrawing from the program, in which case the program would cease to exist. In that event, what would become of the equipment currently at the college?

Epler counseled against distracting ourselves with that possibility yet. He vowed that the college would convert Soekland into a salaried, tenure track employee by November of that year (2001). He advised us to focus our efforts on defining the AS program and getting that curriculum approved.

Thirty students were enrolled in the program during the fall semester of 2001. Working together, people from Spectra, Coherent, Livermore, and Yuba put together a curriculum for an AS degree in optics fabrication, and submitted it for formal approval. But off campus, at the companies, the situation continued to deteriorate. Sales continued to spiral downward. A handful of Yuba students received their certificates at the end of the fall semester, but were unable to find employment.

Early during the spring semester of 2002, Yuba sent an invoice to LEOMA for support during the coming 2002–2003 school year. The companies, now losing money perhaps as fast as they had been making it two years earlier, were in no position to continue underwriting the optics program at Yuba College. On April 12, 2002, I had the unhappy task of writing a formal letter to Yuba, stating that LEOMA and its members, Spectra-Physics and Coherent, were withdrawing their support for the program.

Instruction continued to the end of the 2001–2002 school year, and the program was disbanded in the summer of 2002.

## IRVINE VALLEY COLLEGE

The nascent optical-fabrication program at Irvine Valley College in southern California also fell victim to the collapsing fiber-optics market. Several of LEOMA's larger members, including Newport and Melles-Griot, manufactured precision optics but were too far from Yuba to attract many graduates of that program. Those companies urged LEOMA to establish a program similar to the Yuba program at Irvine Valley College (IVC) in Irvine.

LEOMA held the first organizational meeting at Newport's headquarters in Irvine in December 2000. It was attended by representatives of IVC and more than half a dozen local optics companies, many of them not LEOMA members.[2] LEOMA outlined the program at Yuba and its key components: a curriculum, a classroom, a fully equipped lab, an instructor, and a student-recruiting program.

The curriculum part was easy. Yuba College had already developed an optics-fabrication curriculum that was perfectly suitable for IVC. Classroom and lab space was offered at IVC by Larry DeShazer, director of that school's Center for Applied Competitive Technologies (CACT). A quick survey of companies present in the room that day indicated that much of the necessary lab equipment could be donated by those companies. Nobody could identify an individual suitable to become the program's lead instructor, but that problem and the design of a recruiting program were postponed to future meetings.

But conflicts arose as planning for the IVC program evolved. DeShazer came to an early meeting armed with a long slide presentation that displaced the planned agenda and proposed an approach that was significantly different from LEOMA's. Rather than follow the Yuba template, DeShazer proposed a creating a new curriculum based on a variety of sources, and developing a CACT-centered program that involved less participation by companies than the Yuba program. The meeting ended in uncertainty: Were we following the Yuba template, or the CACT proposal?

Further confusion was introduced in February 2001, when JDSU announced layoffs of optical technicians in San José. In an attempt to get things back on track—one track or another—I invited DeShazer and two key industry representatives to a meeting in one of the airline clubs at John Wayne Airport in Irvine. The industry people felt confident that the JDSU layoffs were a blip. DeShazer emphasized that, unlike the Yuba program, the program at IVC would require no cash inputs from industry. Instead, CACT would obtain grants from the state of California to underwrite the entire program. Moreover, IVC was planning to lease 35,000 sq. ft of space in an empty shopping center, the Irvine Spectrum, and there was ample space for the optics

---

[2]And these nonmembers were not particularly keen on becoming members and paying the associated dues. LEOMA could have required them to become members to participate in the program, but we realized that all these companies had declined to join LEOMA in the past. If we required membership now, most would simply stop participating in the LEOMA/IVC program, but would not hesitate to recruit the program's graduates. Wishing to maximize the program's support by local companies, we made the decision to allow nonmembers to participate along with members.

classroom and laboratory, which would be made available at no cost to industry. The magnitude of industry's contribution to the Yuba program was evolving into a major sticking point in early 2001, so DeShazer's approach was very attractive. We adopted the CACT proposal, but also agreed to use the Yuba curriculum during the initial year at IVC.

The plan was to begin classes at IVC in August 2001. DeShazer was "90% certain" that he could get the funding—approximately a quarter-million dollars—by August 1, and a survey of the equipment available from companies told us the laboratory could be ready by the end of summer.

But by mid-summer the state funding for the program failed to materialize. We postponed the start of classes from August 2001 to January 2002. Nonetheless, CACT remained enthusiastic about the program, and a major hurdle disappeared when a recent retiree from one of the optical companies agreed to become the instructor and act as the program's chief advocate within the IVC system.

We had established an Internet network among the companies involved in the IVC program, which we used to circulate information about curriculum modifications, about stop-gap instructional programs at different companies, and so forth. I remember my startled reaction when the first résumé of an optics technician looking for employment circulated in that network. A year earlier, such a résumé would have had a half-life measured in minutes, and now it was circulating freely, and without takers, on our network.

The state funding again failed to materialize in time for the January 2002 start of classes, and while the official position was that start of classes would be again postponed, until September, 2002, there was diminishing expectation that they would ever start. Over the spring of 2002, as it became evident that the fiber-optics bubble had burst, the companies' interest in the IVC optics program dissolved and disappeared.

## SAN JOSÉ CITY COLLEGE

Unlike the optics programs at Yuba College and Irvine Valley College, the laser-technology program at San José City College, a program that LEOMA boosted to a new life in 2000, continues to this day to graduate skilled laser technicians. The program had existed since the 1970s,[3] but was on the verge of being abandoned in 1999. The lead instructor, John DeLeon, had passed away,

---

[3]I taught there in the late 1970s, and the course I developed there evolved into the *Understanding Laser Technology* course mentioned earlier in this chapter, and eventually into the textbook, *Introduction to Laser Technology*, originally published in 1985 and currently in its fourth edition, published in 2011.

enrollment was anemic, and the curriculum was badly dated. But LEOMA's own survey indicated that laser companies in the San Francisco Bay area alone would need hundreds of new laser technicians in the next few years.

In September 2000, the CEOs of two of the nation's largest laser manufacturers—John Ambroseo of Coherent and Pat Edsell of Spectra-Physics—joined me on a visit to President Chui Tsang of San José City College (SJCC). We showed President Tsang the survey indicating an exploding demand for laser technicians, and described the successful program for optics technicians we had launched at Yuba College. We urged that the laser-tech program at SJCC be rescued, and pledged industry help with revitalizing the curriculum and providing much needed equipment.

Tsang and the dean of Applied Science, Kathy Werle, responded enthusiastically. They agreed that the existing classroom and laboratory space would be retained for the updated laser-tech program. And we set about revitalizing the program. The first organizational meeting, on the SJCC campus, occurred just a few weeks after the meeting with Tsang. Six laser companies attended, and they pledged a total of nearly $200,000 in cash and equipment for the program. We wanted to build a program that would produce both AS graduates from a two-year program, and certificate holders from a one-year program.

The laser portion of the curriculum would be identical for both the one- and two-year programs, but the two-year program would include more math, science, and humanities background. The existing laser-tech courses at SJCC were badly out of date, however, so an entirely new curriculum was necessary. Each of the six companies participated in developing the curriculum, assuring that the graduates would have those skills the companies needed.

But the same problem that hampered the programs at Yuba and IVC was an obstacle at SJCC: the absence of a suitable instructor. We discussed using company employees, on a part-time basis, to fill the gap until a permanent instructor could be found. But I felt it was crucial to begin the program with a full-time instructor, somebody who would "champion" the program through the administrative and financial hurdles it would surely encounter. So we launched a major effort to find an individual who could fill that role.

But toward the end of 2000, an attractive short-term solution presented itself. Sarah Diggs, herself a former laser technician at one of the Bay area companies, had started her own company, BOLT Systems, and had developed what she called a "laser boot camp" for teaching the most basic skills to incoming students. In an intensive, 2-week, 80-hour session, she offered to produce students knowledgeable about handling, cleaning, and aligning optics as well as the basics of laser safety and usage.

So, while still searching for a full-time instructor for the SJCC program, we initiated a series of "boot camps." BOLT Systems provided the instruction, under contract with LEOMA, in the laser classroom at SJCC. During the first six months of 2001, the two-week boot camps were held nearly every month, and the graduates were immediately hired by local companies. Of course, still more training was necessary at the companies. "The boot camps don't produce laser technicians," noted Jim Vargas of Spectra-Physics. "They produce candidates to become laser technicians."

But the companies regarded them as good candidates, and the boot camps were viewed as highly successful. In fact, companies sometimes hired promising but untrained employees, and sent them to SJCC for the boot-camp training. But this arrangement introduced a potential problem, which was solved when all the companies signed letters of agreement not to hire boot-camp graduates who had already been hired by another company.

But the search for a full-time instructor was becoming discouraging. I had joined the dean, Kathy Werle, and two SJCC faculty members on the committee interviewing potential candidates, and we had conducted several unsuccessful interviews. But our luck changed on a chilly January evening in San José. That evening we interviewed Sydney Sukuta, who was finishing his PhD in the Physics Department at the University of Nevada. I remember that, after the interview, the four of us looked at each other and didn't even have to say anything. We had found the right person.

By the time Sukuta accepted a formal offer to join the SJCC faculty for the spring semester, the two-year laser-technology curriculum had been fairly well defined:

Laser 100 covered, over a full semester, the same material that the boot camp covered in two intense weeks. Students who had successfully completed the boot camp and wanted to continue at SJCC could begin with Laser 101.

Laser 101 addressed essentially the same subjects—wave-particle duality, wave optics, and so forth—as the original introductory course in the SJCC laser curriculum.

Laser 102 covered ray and geometric optics, the lens equation, and similar topics.

Laser 103 was based on the second half of my textbook, *Understanding Laser Technology*, and addressed subjects like linewidth control, Q-switching and modelocking, and nonlinear optics.

Sukuta began teaching immediately, in most cases learning the material himself as he taught it. It was a demanding role.

It was a demanding task because Sukuta had almost no hands-on experience with laser technology. He thoroughly understood the underlying physics, of course, but that was a small part of the curriculum. During the summer of 2001, LEOMA and the laser companies sought to remedy this shortcoming by providing Sukuta with eight-week, full-time internships at each company. Sukuta told me recently that those internships were very "valuable in the sense that it affected my delivery of both lectures and labs. My lesson plans were packaged to match what actually goes on in a laser production company from assembly to final tests of lasers and laser systems."

During the summer and fall of 2001, equipment began arriving at the SJCC laser lab. Among the many items was a complete Q-switched, diode-pumped Nd:YAG laser from Spectra-Physics, a fast oscilloscope and precision optical tables from JDSU, and a beam profilometer, IR viewers, and laser-protective goggles from Coherent. Equipment not available from the companies, including computers and software, was purchased.

The program accelerated to its full potential in the fall of 2001, with 46 students enrolled and more than a dozen expected to receive the AS degree in June 2002, and another two dozen in June 2003.

But, as explained in Chapter 1, the economic slowdown of 2001 was affecting LEOMA companies and the LEOMA budget. The fiber-optic bubble had burst, and while that had less effect on laser companies than on fiber-optic companies, the laser companies nonetheless found themselves cutting back. By early 1992, it was clear that LEOMA could not continue operating as it had been, and many significant projects were abandoned. At its May 21, 2002, meeting, the LEOMA board passed a motion that "LEOMA abandon all educational projects from this point forward."

Even so, many local companies continued supporting the San José program individually, and over the years the program has prospered. More than a decade after Sydney Sukuta joined the program, it remains in his capable hands. A handful of fully qualified AS students graduate annually, but perhaps more important, the program provides valuable training for individuals already employed at laser companies. "I have had . . . students with bachelor's, master's, and a few with doctorates, who have taken one or a few of our courses to align with the job market or . . . get a better grip on the hands-on side of things," Sukuta told me.

## THE BENEFIT OF HINDSIGHT

If LEOMA was a solution in search of a problem, educational issues were not a suitable problem. Education was a back-burner issue until it became an emergency, and when it stopped being an emergency, it stopped being an important issue.

The business cycle that moved education from back-burner to emergency to back-burner was not an unusual event, especially in an emerging technology industry like fiber optics. Even with the benefit of hindsight, I don't see a better path than the one LEOMA took. Perhaps education should have been higher among the priorities all along, but the LEOMA companies simply were not willing to commit the resources to do so.

It's interesting that the faculty at San José City College regarded the laser program differently than the faculty at Yuba College regarded the optics program. The faculty at San José City College *wanted* to have the laser program at their school. They were anxious to have me join the faculty committee that interviewed potential instructors, and Sydney Sukuta, the selected instructor, was immediately hired as a full-time faculty member. The school was very cooperative in supplying whatever would facilitate the program. The dean of the college even invited my wife and me to her house for dinner.

At Yuba, one sometimes got the feeling that the college felt the optics program was something it was being forced to do. For reasons that were never clearly explained, it was very difficult to get the instructor onto the Yuba faculty. LEOMA provided nearly a quarter-million dollars in cash, equipment, and time into the optics program, yet a member of the Yuba College Board of Trustees described "corporate greed" as industry's motivation for the program. Perhaps the difference between the two schools was because San José City College is in the middle of Silicon Valley and accustomed to working with industry, and Yuba is in a region where there is very little high-tech industry or employment.

# 5

# EXPORT CONTROLS

## COCOM

Those readers old enough to remember the harshest days of the Cold War—the bomb shelters and the stores of food and water stacked in the basements of public schools—will readily understand the forces that led to the creation of an international group called the Coordinating Committee for Multilateral Export Controls, or "COCOM" for short. Founded in 1949 by the United States and six other nations, COCOM's purpose was to prevent companies in First-World countries from selling militarily critical items to Warsaw Pact countries. COCOM eventually grew to include 17 countries, including those in western Europe, Japan, New Zealand, Australia, Canada, and the United States.

COCOM concerned itself with three classes of commodities: items that could be considered munitions, items related to atomic energy, and items that had both military and commercial applications. Lasers, from their invention in 1960, fell into this latter "dual-use" category, where exports were regulated by the so-called Industrial List. Lasers were controlled by Section 1522 of this list.

One of the first actions of the newly formed Laser Industry Council—LEOMA's earliest forerunner—was a survey of the laser industry to identify the most pressing problems. Export controls were among the top items of the list, and for good reason. The regulations were not only overly restrictive but

*LEOMA and the U.S. Laser Industry: The Good and Bad Moves for Trade Associations in Emerging High-Tech Industries*, First Edition. C. Breck Hitz.
© 2015 by The Institute of Electrical and Electronics Engineers, Inc. Published 2015 by John Wiley & Sons, Inc.

also vague. Several years earlier, an industry executive, Walter Spawr, president of Spawr Optical Research Inc., had been sentenced to 10 years' imprisonment for violating the Industrial List's controls. (Spawr actually served four and a half months in 1983, he told me recently.) Spawr had shipped high-damage-threshold laser mirrors to Germany, but it was unclear whether such lasers were controlled by the existing regulations.

The existing regulations barred the export of items to certain countries that "offer a significant military contribution to a foreign country that could cause a detriment to the national security of the U.S." But Spawr maintained that mirrors similar to his were available from several non-U.S. sources in the international market, and therefore his mirrors were not capable of causing a detriment of U.S. security. Thus, the issue at Spawr's trial, in December 1980, was whether or not his mirrors were controlled by the current regulations.

The jury decided against Spawr, concluding that the mirrors were controlled. But in his appeal, Spawr established that the prosecutor had misrepresented the existing regulations to the judge and jury. Not only was it questionable whether the mirrors were on the control list in the first place, but it was also unclear whether Germany was a proscribed country. "The government attorneys misrepresented the list as a controlling list for all countries," when in fact it was not, Spawr told me.

But the appeals court ruled (2–1) that it didn't really matter what the original jury had been told about the regulations. According to a *Laser Focus* report at the time, the majority held that the Commerce Department's determination alone was "a matter of law, and the government need not establish independently at trial" that the mirrors were, in fact, controlled by existing regulations.

The lone dissenting judge disagreed. Whether or not the mirrors were controlled by the existing regulations was central to the case, the dissent held, and it was an issue that should be determined by a jury, not the Commerce Department.

Spawr pursued the case long after his sentence had been served, eventually appealing to the Supreme Court in 1989. LEOMA joined the Optical Society of America and the International Society for Optical Engineering (SPIE)[1] in filing friend-of-the-court petitions urging the Court to consider the case. Nonetheless, the Court declined the case in October 1989.

The lack of clarity resulting from the Spawr case left exporters very cautious, sometimes to the point of paranoia, about shipping optical goods overseas.

Export controls impacted the industry in other ways as well. In April 1985, the Pentagon had intruded into an SPIE technical conference in Washington,

[1]The SPIE is now known as the International Society for Optics and Photonics.

D.C., barring the presentation of 28 papers in an open session. The papers were unclassified but, according to the Pentagon, contained information restricted by the current export controls. The 28 papers were hastily rescheduled into a special, closed session where all attendees were required to sign nondisclosure agreements. At least one participant refused to sign and was not allowed to enter the room. The affair, viewed by many as a government restriction on the free flow of technical information, caused a great deal of consternation within the laser-optics community.

Spurred by the uncertainty surrounding the Spawr case and the restrictions on conference presentations, the Laser Association of America was trying in the late-1980s to address shortcomings of the export-control regulations. The laser section of the regulations—Section 1522—was a "negative list," in that it listed lasers that were not barred from export. A laser not listed could not be exported without a special license. Negative lists were considered too restrictive, and one of the association's priorities was to convert 1522 to a positive list, listing only those lasers whose export was barred.

In 1987, a labyrinth of committees and agencies dealt with proposed changes to the COCOM export controls. The Electronic Instruments Technical Advisory Committee (EITAC) was a Commerce Department entity, comprised of industry and government representatives. Its role was to collect industry's suggestions and present them to other groups that formulated proposed changes of the COCOM controls.

These other groups included the Defense Department's Technical Work Groups, and the State Department's Technical Task Groups.

The Laser Association focused first on EITAC and engaged Gerry Glen, an EITAC member, as an LAA consultant in early 1987. Meanwhile, I was attempting to join both EITAC and the Defense Department's Technical Work Group 6 (TWG-6, pronounced "twig 6"). My initial application to EITAC was rejected because, as the LAA executive director, I was considered an administrator and "nontechnical." (The powers-that-were either didn't request a résumé, or didn't read the one I submitted; I don't remember which.) That misunderstanding was eventually cleared up and I joined the committee in 1988. Steve Lerner of M/A Com Laser Diode chaired the committee, and shortly after I joined, Larry Cramer of Spectra-Physics also became a member, becoming the third representative of the laser industry on the committee.

The LAA's first action on EITAC was to propose changing 1522 from a negative list to a positive list. EITAC passed the proposal on to TWG-6, where chairman Ed Myers agreed with the concept, and began an attempt to rewrite 1522 as a positive list. Myers was employed by the Institute for Defense Analysis as a defense contractor, and he was perfectly willing to work with industry in formatting 1522 into a positive list. But getting industry representation on his TWG-6 proved to be a thorny issue. The Commerce

Department's licensing officer for lasers, Joe Chuchla, strongly advocated industry participation in the TWGs, but some Defense Department personnel were strongly resistant to the notion. The first time I showed up at a TWG-6 meeting, one of the ranking Defense Department representatives, Al Wall of the Defense Technology Security Agency, walked out and said he wasn't coming back until I left.

Wall was eventually convinced to return to the room, and the process of converting 1522 to a positive list continued. (And, during the subsequent years that we worked together, Wall and I developed a collegial relationship, sharing jokes and later, at COCOM meetings in Paris, dining together.)

In 1988, 1522 was a list of 10 different types of lasers that could be exported without license, provided that they met certain parameters. For example, one paragraph read in part,

(vii) Nd:YAG lasers having an output wavelength of 1.064 micrometres with . . . a pulsed output not exceeding 0.5 joule per pulse . . .

Lasers not mentioned on the list, or lasers whose characteristics didn't precisely conform with those listed, required a license to be exported. Obtaining that license was an arduous, time-consuming, and expensive process.

Unfortunately, there were lasers very similar to Nd:YAG lasers—Nd:YALO lasers in 1988, for example—that could not be exported even with very low output powers. The proposal that LAA put forward in 1988 avoided this problem and simplified the controls on lasers. The LAA proposed a positive list, and that positive list would be based on the characteristics of the output beam, not on what kind of laser produced the beam. The concept was that a laser beam in the 1-$\mu$m spectral range should be controlled at 0.5 J, regardless of what kind of laser produced the beam. Thus, one section of the proposed revision read in part,

I. Pulsed, fixed-wavelength lasers . . .

. . .

    D. [Lasers with wavelength in the region] 1–2 $\mu$m

       1. Everything > 0.5 joule embargoed

This was a positive list in that it described what was embargoed, rather than what wasn't. But, perhaps more importantly, it allowed the free export of any laser whose output was between one and two micrometers, and whose pulse contained less than a half joule. That opened the door to lasers that did not impact the nation's security, but were important to the export trade.

At this point, the proposal to revise 1522 focused on making the control levels clearer and broader, but not changing the levels themselves. The half-joule control in the example cited above was retained in the revision. But as a background activity, LAA was making plans for the next step, after 1522 was revised. A document titled "Lasers and National Security" was being prepared, citing technologies like the Strategic Defense Initiative and laser isotope separation, where lasers played a pivotal role. The document would define thresholds, in terms of power, energy, and wavelength, where national-security concerns became significant.

But the immediate issue was the clarification of the current 1522 control levels. The magnitude of the proposed changes—from a negative list to a positive list based on the properties of the beam rather than the laser—caused consternation in the regulatory agencies. In 1989, each of the three government departments involved in export controls played a distinct role.

The Commerce Department was a fierce advocate of eliminating unnecessary controls. Joe Chuchla, who led the intra-governmental negotiations for Commerce, was often aggressive in "escalating" contentious issues to higher authority, if he thought his adversaries were being unreasonable.

The Defense Department, on the other hand, was leery of the whole concept of positive lists. With a negative list, only those items explicitly approved could be exported; with a positive list, there was always the possibility that a gizmo you hadn't thought of could come along and, because it wasn't described on the list, it could be exported. Often arguing forcefully against various aspects of the revision was Dr. Ray Wick, a civilian employee of the Air Force Weapons Laboratory in Albuquerque.

Standing in the middle acting as a referee between Defense and Commerce was the State Department, which was the official U.S. representative to COCOM. (The Energy Department was also involved in COCOM laser negotiations, but they were not involved in the early intra-government discussions. Energy's concerns were primarily associated with lasers that might be useful in enriching isotopes for nuclear weapons.)

Seeking to further strengthen the support from Commerce, LEOMA[2] organized a visit of industry executives[3] to Washington in the spring of 1989, where we met with Deputy Assistant Secretary of Commerce Jim LeMunyon. It was a successful visit, and as an indirect result, LEOMA was invited to participate in a three-day meeting in Albuquerque among the

[2]The organization's name changed from LAA to LEOMA in May 1989.
[3]The delegation most likely consisted of Glenn Sherman, president of Laser Power Optics, Jon Tompkins, president of Spectra-Physics, and Dean Hodges, a ranking executive at Newport. This is speculation because I simply cannot remember who accompanied me, but these three were heavily involved in the LAA at the time. In recent conversations, neither Sherman nor Tompkins can remember, and Hodges is deceased.

Departments of Commerce, Defense, and State in September of that year. The purpose of the meeting was to discuss the proposal, which had evolved from LAA's submission to EITAC the year before, to convert 1522 from a negative list based on the type of laser, to a positive list based on the properties of the laser beam.

After three days of spirited discussion in Albuquerque, all three departments agreed on a revised form of 1522 to take to COCOM. It was a positive list, but it did not follow LAA's original proposal of basing controls strictly on a beam's spectral content, power, and energy. The problem with that approach is that the type of laser *does* matter. Lasers differ in their efficiency, portability, and ruggedness. For example, in 1989 a solid-state laser producing a multi-watt green beam might weigh a few pounds, require a kilowatt of electrical input, and be rugged enough to bounce around on the seat of a jeep. A gas laser producing approximately the same beam might weigh 10 times as much, consume a dozen times more electrical power, and require rather delicate handling. Clearly, the solid-state laser can be more useful in a military application like defeating sensors or underwater communication.

The U.S. proposal approved in Albuquerque for submission to COCOM was a compromise between controlling the type of laser and controlling the properties of the beam. It defined four broad categories of lasers: gas lasers, solid-state lasers, semiconductor lasers,[4] and liquid lasers. Within each of those four classes, lasers would be controlled by their beam properties.

In December 1989, this proposal went to COCOM, which met in a branch of the U.S Embassy in Paris. For the first time, representatives of industry were included in the U.S. negotiating team to COCOM, and I participated in the Paris discussions representing LEOMA. The meetings lasted four days, with all conversations taking place in French with immediate English translation, or vice versa. The U.S. proposal was discussed extensively during those four days, but no conclusion was reached.

The global political situation was changing rapidly in 1989, and those changes were affecting COCOM. Relations between the United States and eastern European nations were thawing as those nations distanced themselves from Russia and the Soviet Union. A front-page article in the December 17 issue of the *New York Times* reported that the (first) Bush administration was planning a significant relaxation of the country's export controls.

But laser controls were not among those initially targeted for relaxation by the Bush administration, and it would be another two years before the Soviet

---

[4]Semiconductor lasers are, of course, solid-state devices. However, they are so different from conventional, optically pumped solid-state lasers that they are considered a class of their own, and the term "solid-state laser" is broadly understood to exclude semiconductor lasers.

Union officially dissolved. So the COCOM controls were still of vital interest to the U.S. laser industry.

In March 1990, LEOMA held its first executive seminar in Monterey, California. One of the speakers was Jim LeMunyon, the Commerce Department official we had visited the previous year. LeMunyon used the occasion to make an official announcement of the new "GCT license," a general license that would allow export of controlled items to other COCOM countries. This license greatly alleviated the difficulties of laser companies, since the majority of their overseas sales were to customers in other COCOM countries. Two decades later, Hank Gauthier, Coherent's long-time CEO, would recall the GCT license as LEOMA's greatest accomplishment during his years with the association.

The GCT license, together with the pending revision of 1522, promised to virtually eliminate the burden of export controls on U.S. laser manufacturers. "LEOMA would like to take credit for this," wryly noted LEOMA secretary Bob Gelber[5] in the minutes of the May 1990 board meeting, "but world history probably gets a footnote."

It still remained, however, to guide the revision of 1522 through COCOM, and that was of sufficient interest to laser manufacturers that the laser representation on EITAC had grown to five members. Joe Chuchla of the Commerce Department created a separate committee, LaserTAC (Laser Technical Advisory Committee) to address COCOM's treatment of 1522. I was named the committee chair, and the committee included Rich Hooper of United Technology Research Labs, Steve Lerner of M/A COM Laser Diode, Larry Cramer of Spectra-Physics, and Don Siebert of Allied-Signal.

LaserTAC, together with the TWG (Defense) and TTG (State) committees, analyzed the input about 1522 that other countries had put forth at the December COCOM session. A bilateral meeting with the U.S. and German COCOM delegations was held at the State Department in Washington in the spring of 1990, and from that meeting emerged a consensus that we carried to COCOM in June 1990. The U.S. and German delegations were successful in convincing other COCOM countries that lasers should be controlled within four broad categories (solid, liquid, gas, and semiconductor), and within those categories controls should be based on the beam properties. The November 1990 issue of LEOMA Newsletter reported that the new 1522 had been published in the *Federal Register* and was now official.

There was another piece of good news. LEOMA's document, "Lasers and National Security," had precipitated discussion about the levels of laser controls. Although the original intention in revising 1522 had been simply to enhance its understandability, along the way there had also been significant

---

[5]Gelber was an executive with Coherent.

**TABLE 5.1    A Comparison of Control Levels Before and After the COCOM Revision of 1990**

|                  | Old 1522        | New 1522              |
| ---------------- | --------------- | --------------------- |
| $CO_2$           | 5 kW            | 10 kW                 |
| Ti:Sapphire      | Embargoed       | 20 W average power    |
| Cw Nd:YAG        | 50 W            | 500 W                 |
| Excimer          | 120 W average   | 500 W average         |
| Nonlinear optics | Embargoed       | 2xYAG ok up to 30 W   |
| Dye lasers       | 1 W             | 20 W                  |
| Ar/Kr ion        | 20 W            | 30 W                  |

relaxation of the levels of control. According to the LEOMA newsletter cited in the previous paragraph, the powers at which common lasers were embargoed had risen as indicated in Table 5.1.

Meanwhile, world history continued to evolve, and the British were pressing to substitute a "Core List" for the existing COCOM controls. This list would define those commodities that were essential to the core security of COCOM nations, and allow everything else to be exported. There was much ado about this proposal at COCOM, but it was eventually accepted. The effect of the Core List on laser controls, however, was nonexistent. Since 1522 had just been revised in the preceding months, 1522 was imported wholesale into the new Core List.

In the years while 1522 was being revised, another issue was working its way through the U.S. export-control bureaucracy. The controls on optical components in the 1980s were vague at best (witness the Spawr case), and pretty much all inclusive. Especially after Spawr was convicted, optics manufacturers took pains to obtain a license for pretty much any questionable component they exported. But the emphasis on clear, positive lists that was forming in the late 1980s brought pressure to clarify the controls on optical components, and to revise those controls to make them a positive list.

What were the pressing national security concerns with optical components—seemingly harmless items like mirrors, polarizers, and so forth? The Defense Department was rightly concerned that these components could be used to generate and manipulate high-power laser beams, beams perhaps powerful enough to be employed as weapons. So a natural parameter to control would be a component's damage threshold. If an optical component suffered damage when exposed to a relatively benign power density, it would not be useful with a high-power laser. So the Defense Department proposed to control optical components based on their damage thresholds.

That proposal brought howls of protest from optics companies, including many LEOMA companies. Damage thresholds were difficult to measure accurately, and there were only a few laboratories in the country that could make such measurements. The measurements were expensive, and the capacity of the existing labs was no match for the volume of optics that companies wanted to export. Controls on optical components were discussed heatedly at the Albuquerque meeting in September 1989, but no firm conclusions were reached.

Discussions between LEOMA and government agencies continued during the next year until, happily, the government was convinced that controls based on damage thresholds were not optimal. The Defense Department agreed that the U.S. proposal going to COCOM in December 1990 would embargo only optics "explicitly designed for multi-kilowatt lasers." Although that phrase lacked the precision many people desired, it was better than the vague ambiguity that had been in effect during the previous decade.

But further trouble awaited at COCOM. Several nations, led by the United Kingdom, insisted that such high-power optics should not be controlled by COCOM's Core List at all, but should instead be placed on the International Munitions List (IML). The United States opposed this because there was no firm agreement among nations about how to treat items on the IML. In some cases, it was felt that placing high-power optics on the IML would actually make them *easier* to export, a result that the U.S. Defense Department strongly opposed.

Thus, shortly after the *Federal Register* announced the successful revision of laser controls under 1522 in the fall of 1990, the LEOMA Newsletter of March 1991 bemoaned the situation with optics controls:

> The lack of agreement within COCOM leaves the current situation ambiguous, and it is not at all clear which optical components now require export license and which do not. This delicate and disturbing situation will require careful monitoring during the coming months.

With COCOM unable to reach agreement, the discussion of optics controls shifted to a higher diplomatic level during the summer of 1991, where an accord was finally achieved. The agreement placed optics controls on COCOM's Core List and not on the International Munitions List. Moreover, those controls were based on parameters other than damage thresholds. These new controls were complex, but less vague and less restrictive than the previous negative controls. They controlled specific items, like large ZnSe optics, lightweight mirrors, and space-qualified optics, with clear definitions of all controlled items.

So, by the end of 1991, the export controls on both lasers and optics— controls that had caused manufacturers so much difficulty in the 1980s—had been reformed. In the official report of LaserTAC to the Commerce

Department in September 1992, I wrote (perhaps exaggerating a little boastfully), "U.S. laser and optics manufacturers have probably experienced less difficulty with export-control regulations during the past year than during any year since the laser was invented."

Despite that boast, there were still issues to contend with. In 1991, Dirk Kuizenga of LaserScope published a paper describing his 100-W green laser, which used nonlinear optics to convert the beam of a 150-W infrared laser into the green spectral range. The 150-W infrared lasers could be exported without license, and the nonlinear crystal that converted the infrared output to green could be purchased from several vendors, including several in China.

So the Defense Department, envisioning enemies equipped with pilot-dazzling green lasers proliferating around the world, demanded a "rollback" of the control on infrared lasers. But 150-W infrared lasers were manufactured by the dozens by lasermakers around the world, and they were crucial in many industrial applications. Reluctantly, the Defense Department acknowledged that a rollback was not feasible. Years later, I was told by a Defense Department official in position to know, that the cockpit of virtually every military aircraft in Iraq and Afghanistan was equipped with goggles to protect against green lasers.

Despite this and other minor issues, manufacturers continued to experience little difficulty with export controls for most of the next decade. The Soviet Union dissolved in December 1991, and COCOM went out of business on March 31, 1994. There was no further need for LaserTAC, and it simply ceased to exist. The United States and other countries continued to maintain unilateral export controls, but these controls did not significantly impede exports of U.S. laser or optics manufacturers.

## WASSENAAR

But global political issues continued to evolve. There were emerging national security concerns about the destabilizing effect of selling militarily significant commodities to Third-World countries whose long-term goals were dubious. Diplomats meeting in the Dutch town of Wassenaar in July 1996 agreed to a new regime of internationally coordinated export controls, and the so-called Wassenaar Arrangement was born. Signatories included most industrialized countries, Russia, and several other former members of the Soviet Union, but China, Israel, and most Third-World countries were excluded.

For lasers and optical components, the Wassenaar Arrangement adopted the former COCOM controls almost verbatim. Some new controls were added to deal with diode lasers, which had become important in recent years, but these controls were of little concern to LEOMA members. In October 1998, the Defense Department's Laser and Optics Technical Work Group, still chaired by Ed Myers, began a review of the laser controls. Myers, familiar with LEOMA from the days of COCOM, invited LEOMA to participate. I attended the meeting, where the focus was on high-power lasers, but the controls proposed there affected such high-power devices that they were of little concern to LEOMA.

But as calendars for the last year of the century began appearing on walls and desks, laser export-control issues started being of more concern in the United States. The Defense Department had undertaken a very thorough study of the military significance of lasers and their availability throughout the world. The study was led by Ray Wick, who was now retired from the Air Force Weapons Lab, and his colleague John McMahon, recently retired from the Naval Research Labs. Both were respected and competent laser physicists, now working as consultants for DOD contractors. The study was to form the backbone of the laser section of the Militarily Critical Technologies List, which would in turn become DOD's guide for formulating export controls at Wassenaar.

In September 2001, McMahon published a report, "Military Requirements for Export Controls of Lasers During the Next Decade," a 114-page volume that would set the stage for the coming debate.

The Commerce Department's EITAC committee had been dissolved with the end of COCOM, but as export controls in general became of greater concern to U.S. industry, the Commerce Department created a new committee. The Scientific Instrumentation Technical Advisor Committee (SITAC) was to perform the same function that EITAC had performed: provide U.S. industry with a mechanism to influence export controls. I applied for SITAC membership in October 2001.

During 2002, the Defense Department's Lasers, Optics, and Imaging Technical Work Group (LOITWG), now chaired by Ray Wick, worked on finalizing its revisions to the Militarily Critical Technologies List (MCTL). Wick invited industry participation in the meetings, and several people— among them Bill Wells of Spectra-Physics, Merrill Apter and Paul Rudy of Coherent, and I—represented laser manufacturers in at least some of the LOITWG meetings. LOITWG was a Defense Department committee, however, and the MCTL was a Defense Department document. Wick retained the authority to disregard industry input in the committee's final decisions, and in the end he did just that in every case. The MCTL that emerged from his committee in 2003 reflected the Defense Department's perspective on all issues that had been debated in the committee.

Next, the Defense Department set about creating a U.S. proposal to Wassenaar that would embody many of the new MCTL control levels.

That entailed negotiations between Defense and Commerce but, unlike the situation during the COCOM years, these negotiations were conducted without significant industry participation. Rather than being an advocate for industry, it seemed to me at the time that Commerce had decided it was on the government's side in what it viewed as a debate between government and industry. In the COCOM days, the Commerce Department was so appreciative of industry input that it picked up my airfare to SITAC meetings in Washington, and even to COCOM sessions in Paris. In the twenty-first century, industry pays for everything.

An example of the changed attitude was the proposed control on semi-conductor lasers that Defense forwarded to Commerce in mid-December 2002. The proposed controls were significantly more restrictive than the then-current controls. Commerce waited until Thursday, January 23, to e-mail the proposal to SITAC members, and requested response by Monday, January 27. This allowed no time for thoughtful response, much less any sort of coordinated response from industry.

This was a different dynamic than the one a decade earlier, when Joe Chuchla and others at Commerce bent over backward to ensure that industry's perspective was represented. Seeking to restore the balance, LEOMA arranged a meeting with Commerce Undersecretary Kenneth Juster, who directed Commerce's activities in export controls. We were hopeful, based on our experiences in previous visits to ranking Commerce officials, that our visit would result in a more cooperative relationship between LEOMA and Commerce.

Coherent's president John Ambroseo and Spectra-Physics' president Guy Broadbent joined me on the visit on March 3, 2003. We met with Under-secretary Juster and Assistant Secretary James Jochum. The meeting was "short and perfunctory" and the officials gave little indication of concern for our views, according to my notes from the time. I recall being embarrassed that I'd wasted Ambroseo's and Broadbent's time by urging them to participate in the meeting.

After visiting Commerce, Ambroseo and I visited the Defense Technology Security Administration (DTSA), where we met with Sam Sevier, the administration's deputy director and several other DTSA officials, including Woody Anderson, who was to be a crucial participant in the laser discussions during the ensuing years. The meeting with DTSA was quite constructive, with Sevier emphatically requesting input from industry during the coming revision of laser export controls.

So in March 2003, we were left with a somewhat ironic situation. We had what appeared to be a cooperative and constructive relationship with our traditional adversaries at DTSA, and we had an at-best ambiguous relationship with our traditional advocates at Commerce.

Meanwhile, technology was catching up with the laser controls that had seemed so unrestrictive when COCOM passed them in 1990. Lasers deemed exotically powerful in 1990 were more commonplace in 2003. At the same time Defense was pushing for stricter controls, industry was clamoring for a relaxation of the existing controls.

In May 2003, LEOMA submitted to SITAC a proposal for a significant revision of the Wassenaar controls on lasers. It was a three-column spreadsheet, one column showing the current controls, one showing the LOITWIG recommendations, and the third showing the LEOMA recommendations. That spreadsheet was to become the central instrument of intra-U.S. discussions for the next several months.

SITAC deals with scientific instruments that have both military and civilian uses. That includes lasers and other instruments, but night-vision devices are by far SITAC's greatest concern. Because most SITAC members had little experience with lasers, and the laser representatives on SITAC had nothing to contribute regarding night vision, a separate subcommittee was formed to address laser export controls. The Laser Working Group had two goals: to formulate a final industry position and to reconcile that position—to the extent possible—with the Defense position.

Informal discussions and phone calls among industry members during the spring of 2003 led to a formal meeting of the Laser Work Group in July at Coherent in Santa Clara, California. Wayne Hovis, the laser licensing officer at the Commerce Department, along with scientists and engineers from six or seven laser manufacturers, participated in the meeting. The agreement that emerged from that meeting met the first goal: it was a formal proposal of the export controls advocated by laser manufacturers.

The next step was reconciling that proposal with the position of the Defense Department. There were numerous issues on which Defense and industry saw things differently, but I will describe two as representative of many more. One of these issues was solved in the early 2000s, and the other remains unresolved today.

Everybody knows that lasers produce intense beams of light that are extremely monochromatic. But a technology called nonlinear optics can change the color, or wavelength, of a laser's beam. A nonlinear crystal placed in the beam of an infrared laser, for example, can convert the invisible beam to a green beam. The problem is that the Wassenaar rules explicitly control *lasers*. The word is an acronym for "light amplification by stimulated emission of radiation." Nonlinear optics does not involve "stimulated emission of radiation," yet it is nonlinear optics that generates the green beam in the example. So: Is the green beam controlled by the existing controls on lasers?

The second issue, one that remains an issue today, concerns how fast a laser beam spreads as it travels. The more it spreads—or diverges—the lower the power density on its target. Clearly, a highly divergent beam is of less concern than a well-collimated beam. The question was how to quantify a beam's divergence. The original COCOM controls simply put more stringent limits on a "single-transverse-mode" laser than a "multimode" laser. Laser engineers understand what the divergence of a single-transverse-mode laser is, but licensing officers do not and that term was undefined in the Wassenaar controls of 2003. As a result, different licensing officers in different countries applied different controls to lasers exported from their countries.

The first issue wasn't so much a difference between industry and Defense as it was difficulty finding the correct wording. Everybody knew what we wanted to mean, but we couldn't agree on how to say it.

The second issue was more contentious. The International Organization for Standards had created a standard[6] for measuring the divergence of a laser's beam. (LEOMA's involvement with the ISO laser standards is discussed in Chapter 3.) LEOMA advocated that a reference to that standard replace the terms "single transverse mode" and "multimode" in the Wassenaar controls.

At a meeting in Washington on August 22, 2003, among Wayne Hovis of Commerce, Woody Anderson and John McMahon of Defense, and me, McMahon argued that the ISO standard did not work with many kinds of laser beams, and he formalized his argument in a position paper issued in November. He held that using the ISO standard would result in many single-mode lasers being classified multimode, and hence exportable without license.

LEOMA countered McMahon's argument by pointing out that the lasers he cited—semiconductor lasers and unstable-resonator lasers, for example—had highly divergent beams and could never be considered "single transverse mode" under any circumstances.

Discussions—face-to-face meetings, e-mails, and telephone calls—of these and other issues continued through the winter of 2003–2004. One important meeting, attended by Woody Anderson and Al Paxton of Defense, Wayne Hovis of Commerce, and more than a dozen laser manufacturers, took place at a major laser conference in San José, California, in late January.

Many issues were resolved at these meetings, and some were not. Among those resolved was the question of nonlinear optics. The U.S. proposal to Wassenaar would include a note, "The coherent optical output from any system containing a laser, whether the output arises directly from stimulated emission or is converted by intracavity or external nonlinear optical devices, is subject to control under 6A005."[7]

And among the unresolved issues was the question of beam divergence. McMahon had come to agree that the parameter defined by the ISO standard to

[6]ISO 11146.
[7]"6A005" identifies the main section of the Wassenaar controls that applies to lasers.

measure beam divergence or quality—the so-called M-square parameter—was an appropriate parameter for export controls. But now Defense insisted that any laser with an M-squared value less than two should be considered "single transverse mode." Industry refused to accept this interpretation because it would subject many multimode lasers to the stricter controls intended for single-mode lasers.[8] As a result, the U.S. proposal to Wassenaar retained the undefined phrase, "single transverse mode." As of the writing of this book, that phrase remains in the Wassenaar controls . . . and it is still undefined.

LEOMA presented the U.S. laser proposal to the Wassenaar Experts' Group session in Vienna on April 21, 2004. But Wassenaar in 2004 was a different animal than COCOM had been in 1989. At COCOM, maybe half the people in the room were at least familiar with lasers, and several were world-class experts. Heinrich Endert and Dirk Basting of the German delegation, Kunihiko Washio of the Japanese delegation, and others whose names no longer come to mind had spent most of their careers working with lasers. In the Wassenaar Experts' Group in 2004, I don't believe there was anyone outside of the U.S. delegation who was remotely familiar with the technology of lasers. The delegates were all diplomats—intelligent, astute, and generally well-informed individuals, but not knowledgeable about lasers.

After LEOMA presented the U.S. proposal, there were several comments from around the room, but they were comments prepared in advance of the meeting. The U.S. proposal called for an embargo of green lasers whose output was created by nonlinear optics, at 30 W. One nation announced that until now, these lasers hadn't been controlled at all. (That nation held that lasers whose output was generated with nonlinear optics were not controlled by the existing Wassenaar laser controls.) Because they had not been controlled previously, the nation argued they should remain uncontrolled or, at most, be controlled at 200 W.

Even more problematic was the issue of fiber lasers. These lasers, consisting of a thin strand of glass many meters in length, had not existed outside of a few laboratories in the COCOM days, and were usually considered subject to the normal Wassenaar controls on solid-state lasers. But these lasers were (and are) enjoying enormous commercial success because of their *differences* from other solid-state lasers: they were more energy efficient, more compact and rugged, and produced less-divergent beams than conventional solid-state lasers. These advantages also made them much more suitable for military

---

[8]Speaking very rigorously, a laser truly oscillating in a single transverse mode will have an $M^2$ exactly equal to 1. In practice, it's virtually impossible to create a commercial laser with a measured $M^2$ of 1, and lasers with an $M^2$ less than about 1.3 are generally considered to be single-transverse-mode lasers. But lasers with $M^2$ between 1.3 and 2 are multimode, and accepting the Defense position would have subjected these lasers to the stricter single-mode controls.

applications, and the U.S. proposal advocated subjecting them to stricter controls than conventional lasers.

But in the spring of 2004, the U.S. proposal on fiber lasers was met with a perplexing comment by one of the Wassenaar delegations. That nation contended that a fiber laser was not a solid-state laser, and therefore uncontrolled and should remain so. But a thin glass fiber is surely a solid-state material, so the statement that "a fiber laser is not a solid-state laser" was rather confounding.

But aside from these few comments, there was little discussion of the U.S. proposal at the spring Wassenaar session of 2004. The proposal was put on "study-reserve," which meant that the respective Wassenaar nations would take the proposal home to their experts, and return in the autumn with comments and counterproposals. We were hopeful that the proposal could be modified to accommodate the other nations' inputs, and officially approved at the fall meeting.

It didn't happen. The issues were too complex, and while the U.S. delegation had the resources at the table to answer questions raised at the fall Experts' Group meeting, the diplomats posing the questions lacked the resources to process the answers. Unable to produce any fruitful discussion, the fall 2004 session ended with several nations putting the laser proposal on study-reserve.

There were several surprises at the spring 2005 Experts' Group session. The nation demanding that green lasers be exportable at powers up to 200 W was represented by different individuals . . . and the 200-W demand disappeared. Also, the nation that had insisted, "A fiber laser is not a solid-state laser," suddenly dropped that assertion. So, two major stumbling blocks from the previous year had simply vaporized.

But the biggest surprise was when Nolandia,[9] largely silent about lasers until now, brought a completely new proposal to Vienna. Unlike the U.S. proposal, the Nolandian proposal based controls on wavelength and beam power, not on the type of laser producing the beam. This was the same concept LEOMA had proposed in 1988, but had ultimately abandoned as impractical. And there were several glaring loopholes on the Nolandian proposal, including the omission of any mention of nonlinear optics. Nonetheless, the U.S. delegation decided to support the Nolandian proposal.

I explained earlier that Wassenaar was a different animal than COCOM had been. That was true because there was significantly less laser expertise at Wassenaar, but there was another factor. To the best of my recall, proposals at COCOM from any nation were evaluated at face value. At Wassenaar, the

---

[9]Wishing to follow Wassenaar protocol and not identify Wassenaar nations with their positions on various issues, I have created "Nolandia."

United States had earned itself a reputation as something of a bully because it had engaged in some intense arm-twisting in the past to impose strict controls on many commodities. The U.S. invasion of Iraq in 2003 had not lessened that image. As a result, U.S. proposals at Wassenaar were greeted with a heavy dose of skepticism.

Thus, the U.S. delegation to Wassenaar decided to scrap its own laser proposal and embrace the Nolandian proposal, because a "Nolandian proposal" had a greater chance of success at Wassenaar than an "American proposal."

Because the proposal was still very complex, a laser technical work group was formed of nations concerned with the laser controls, and chaired by Nolandia. The work group met during a recess of the Experts' Group formal session, but lacked the expertise to move ahead. Recognizing the need for laser expertise, the Wassenaar chair directed that the laser work group meet during the summer of 2005, and requested that concerned nations ensure that experts were present.

And there was still one more surprise before the spring meeting was over. On the last day of the spring session, the Nolandian delegate announced that he had received instructions from his capital to relinquish the chair of the laser work group, because the Nolandian delegation to Wassenaar lacked sufficient resources. The chairman asked for another nation to volunteer to chair the work group.

There were 50 people in the room, seated around the large, hollow, rectangular table. Silence prevailed. After 60 seconds, the chairman repeated his request. Again, silence. After another 60 seconds, the head of the U.S. delegation to Wassenaar reluctantly tilted the U.S. nameplate to an upright position, accepting the chair of the laser work group. The diplomatic advantage of having a chair from anyplace but the United States had been lost.

The special session of the Wassenaar laser work group in the summer of 2005 was a surprisingly productive meeting. Several nations sent knowledgeable laser experts, all of whom focused on the Nolandian proposal. As it was expressed colloquially that summer, the concept of the proposal was to control what comes out of the box—that is the laser beam—instead of what's in the box. As LEOMA had discovered in 1988, the flaw in that approach is that what's in the box can matter. Early in the summer of 2005, John McMahon explained the situation very succinctly in an essay entitled, "Sometimes Size Does Matter." Size, weight, electrical efficiency, requirements for cooling water, ruggedness, beam quality—all these factors affect the military usefulness of a laser, and they all depend on what's in the box.

Nonetheless, the United States supported the Nolandian proposal and during that summer session in Vienna, the experts struggled to modify it so it would be acceptable to all parties. The result was a two-part proposal to be presented at the autumn session of the Wassenaar Experts' Group. The short-term proposal contained a clarification of the control of nonlinear optics, unambiguously controlling lasers whose output was created by a nonlinear process. Another part in the short-term component was an agreement on how to control fiber lasers, whose status had been unclear in previous controls. The expectation was that the short-term proposal could be formally accepted at the fall meeting.

The long-term component that emerged from the summer session of 2005 was a complete revision of the Wassenaar laser controls. Based loosely on the original Nolandian proposal, it would control most lasers based on the characteristics of their output beam, but identified a handful of specific lasers that were to be controlled separately.

Both components were presented at the fall Experts' Group meeting as the "Nolandian-American proposal." One country placed the short-term component on study-reserve, but removed that constraint in the weeks following the session. So, for the first time in two years, a successful modification of the Wassenaar laser controls had worked its way through the system.

The long-term component was a complex, eight-page document that tended to glaze the eyes of even the experts who had crafted it that summer, and was surely unfathomable to the non-laser-experts in the Experts' Group. In an attempt to facilitate explanation, LEOMA created a series of charts that presented a more-comprehensible visual representation of the long-term proposal (Figure 5.1). As expected, the long-term component was put on study-reserve, and the delegates took the proposal, along with the LEOMA charts, to their respective capitals for further study.

There were three meetings of the Wassenaar laser work group in 2006, two of them during recesses at the Experts' Group meeting, and another special session in the summer of that year. The LEOMA charts were repeatedly modified as the proposed controls were nudged this way and that in the discussions. But the overall concept of the controls reflected the philosophy of controlling what comes out of the box, rather than what's in the box.

But embracing that philosophy required several seemingly extreme contortions. One example is pulsed solid-state lasers. These lasers can be pulsed with an internal optical gate (called a Q-switch) that lets the light out in bursts rather than in a steady stream. But the electrical input to the laser can be supplied in two ways: either continuously, or in pulses timed to be synchronous with the Q-switch. The two approaches produce very different lasers, and in the old Wassenaar controls that difference was recognized by having different controls on lasers that were "continuously excited" from those that were "pulse excited." But that technique was rejected in the new controls, because it was controlling what was in the box, not what came out of the box.

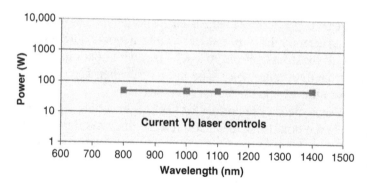

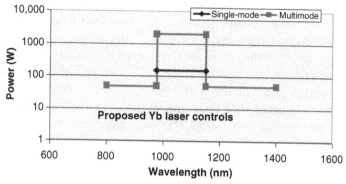

**FIGURE 5.1** LEOMA produced a series of charts, like this one, to explain to Wassenaar delegates the difference between existing controls and the controls in the "Nolandian-American proposal." In this case, ytterbium lasers were controlled at 80 W under the existing controls. Under the proposal, the controls were significantly relaxed, for both single-mode and multimode lasers, in the spectral range around 1000 nm. These two simple charts represented the information contained in half a page of mind-numbing subparagraphs and sub-subparagraphs in the actual documents.

Instead, paragraph 6A005.b.6.b.1.b. of the Wassenaar controls reads as follows: "[Any solid-state laser having] Average power exceeding 20 W *limited by design to a maximum pulse frequency less than or equal to 1 kHz* [requires a license]." The phrase[10] in italics is code for "This paragraph applies only to lasers that are pulse excited." It's code that is comprehensible only to the most nerdy laser techie, and its purpose is certainly less than clear to licensing officers and other administrative officials. But it controls what comes out of the box.

The most contentious issue, and the last to be resolved, involved fiber lasers. The United States felt there were national-security issues, and presented an essay, with plentiful charts and calculations, showing the havoc that

---

[10]Full disclosure: I am the author of that abominable phrase. There was no other way to comply with the control "what-comes-out-of-the-box" requirement.

could be wreaked by a terrorist with an easily portable, 100-W fiber laser. The essay was sufficiently scary to convince other nations, normally resistant to U.S. proposals for strict controls, to agree that fiber lasers should be subject to tighter controls than other solid-state lasers. But then came the problem of conforming to the what-comes-out-of-the-box rule.

Having a separate set of controls for fiber lasers would clearly be controlling what's in the box. Many delegates felt the rule would have to be bent in this case, but the delegate from one nation insisted that the rule be observed. My personal intuition was that the delegate resisting a solution was smarting at having acquiesced to U.S. insistence on stricter controls for fiber lasers, and was now determined not to acquiesce to a U.S. position again.

Finally, at the third and final laser session of the final week, a solution was reached. As a result, paragraph 6A005.a.6.b.1. of the Wassenaar controls now reads: ("[Any laser having] Wall-plug efficiency exceeding 18% and output power exceeding 500 W [requires a license]." The wall-plug efficiency of a laser is the ratio of power in the laser's beam to the electrical power the laser takes from its plug in the wall. Only fiber lasers[11] are likely to have wall-plug efficiencies in excess of 18%, so the efficiency requirement is code for "This paragraph applies only to fiber lasers." But again, it is code that is doubtlessly unclear to many licensing officers and administrative officials.

Finally, in a nod to the manufacturers of massive industrial lasers whose sheer weight made them impractical for military applications, a note was added: "6A005.a.6.b. does not control . . . industrial lasers . . . with a total mass greater than 1,200 kg" (see Figure 5.2).

But even the most ardent supporters of the what-comes-out-of-the-box rule had bent the rule in some cases, so at the end of the day, six types[12] of lasers were explicitly controlled by what's in the box rather than what comes out. Still, the new Wassenaar laser controls achieved in 2006 differed philosophically from the previous controls, as outlined in Table 5.2.

These adjustments finally satisfied the objections of all delegations, and the newly revised laser controls were unanimously approved by the Experts' Group. It had taken three years of discussion at Wassenaar. A spontaneous round of applause broke out around the large, rectangular table.

From 2006 until the publication date of this book, there have been few changes in the Wassenaar controls of lasers. Some modifications were made for semiconductor lasers, and in 2010 the Obama administration proposed revamping all the country's export controls into a tiered system. For a short

---

[11]Semiconductor lasers and $CO_2$ lasers can have efficiencies over 18%, but both these lasers are controlled under the "Other lasers" section of the new 6A005.

[12]Diode lasers, carbon mon- and dioxide lasers, excimer lasers, chemical lasers, and non-repetitively pulsed glass lasers.

**FIGURE 5.2**   Struck by the seeming silliness of the efficiency and weight controls imposed on fiber lasers, Nufern—a manufacturer of the fibers used in fiber lasers—presented a concept of what techniques a fiber-laser manufacturer might employ to avoid the Wassenaar controls. The bricks in the cabinet push the laser over the 1200-kg limit, and the electrical toasters and heaters on top reduce its wall-plug efficiency to less than 18%.

**TABLE 5.2**   **The Revised Laser Controls of 2006 Reflected the Philosophy of the "Nolandian-American Proposal" to Control Lasers by External Parameters Rather Than by Type of Laser**

| Old 6A005 | New 6A005 |
|---|---|
| a. Gas lasers | a. Nontunable continuous lasers (divided into eight categories by wavelength) |
| b. Semiconductor lasers | b. Nontunable pulsed lasers (divided into eight categories by wavelength) |
| c. Solid-state lasers | c. Tunable lasers (divided into three categories by wavelength) |
| d. Dye and liquid lasers | d. Other lasers (six types of lasers controlled separately) |

time there was concern that lasers would be included in the most-restrictive tier, and in early 2011 LEOMA organized a visit by half a dozen industry executives to the Commerce Department, where we had a productive meeting with Assistant Secretary Kevin Wolf and other Commerce officials. They assured us that lasers would not be placed in the most-restrictive tier, and the current thinking is that the Obama administration's revisions of export controls will have little impact on the laser controls.

## THE BENEFIT OF HINDSIGHT

It's easy to underestimate the difficulty in inducing a modification of the U.S. export controls for dual-use commodities. Several years ago, a laser manufacturer wanted to manufacture its lasers in China, but part of the technology involved was subject to Wassenaar controls. The company estimated it could cut costs by tens of thousands of dollars a year by manufacturing in China, so it was an important issue for them. They requested time to make a presentation at SITAC, and sent several engineers to make the case. Their presentation was well received at SITAC . . . but the story ended there.

Making a SITAC presentation is the first step in changing the controls, but it's important to follow up and see that the proposal is formalized and submitted to interdepartment discussions. Visits to Commerce officials can be helpful. Formal membership on SITAC can be helpful. Visits to Defense officials can be helpful. If, after a process that can take many months, a proposal is accepted by the Departments of Commerce, Defense, and State, it will be submitted to Wassenaar. At that point, it can be very helpful if the proposal's original sponsor can get himself appointed as a member of the U.S. delegation to Wassenaar, so he can be present when compromises and modifications are made there.

# 6

# THE FEDERAL GOVERNMENT

LEOMA's most significant interaction with the federal government, twice undertaking successful revisions of the nation's export control laws, is described in the previous chapter. But the federal government and its policies can have enormous impact on industry beyond export controls, and as early as November 1988, the LAA board directed me to identify new ways the association might constructively interact with the government. But I didn't make any strong recommendations, because we had our hands quite full with international standards and export controls, and even if there had been new paths to constructive interaction, I doubt we could have explored them.

By 1991, LEOMA had made significant headway with the major issues— export controls, standards, and conference proliferation—that had dominated its earliest years. The association began seriously examining other areas where companies might work together for the common good. In March of that year, I circulated to the board a recommendation that LEOMA address the industry's relationship with the federal government.

The recommendation pointed out that LEOMA and its members had little interaction with the federal government. As a result, they missed out on many opportunities, such as a million-dollar grant from the Commerce Department to a non-LEOMA company to develop diode lasers. The recommendation further cited potential benefits of improving the interface with the federal government, for example, financial support of our work in international

*LEOMA and the U.S. Laser Industry: The Good and Bad Moves for Trade Associations in Emerging High-Tech Industries*, First Edition. C. Breck Hitz.
© 2015 by The Institute of Electrical and Electronics Engineers, Inc. Published 2015 by John Wiley & Sons, Inc.

standards, improved access to European markets, and further easing of export controls.

At its May 1991, meeting the LEOMA board accepted the recommendation and identified "development of government–industry relations" as a new goal of the association. One of the first opportunities to implement that decision occurred shortly thereafter, when the Defense Department undertook a survey of the U.S. industry to identify the strengths and weaknesses that could affect national security. LEOMA coordinated input from the U.S. laser industry so that it presented a coherent and accurate picture of the industry's capability.

Industry data in the 1990s were collected in categories defined by a system called the Standard Industrial Classification (SIC) System. The SIC was useful to many segments of the U.S. manufacturing industry, because it provided informative market data. But the SIC data were of little use to laser manufacturers, because lasers were lumped into a miscellaneous SIC category that included Christmas tree lights and electric outboard motors, among other things. The SIC System was administered by a branch of the Commerce Department, the Census Bureau, and in 1997 the Census Bureau was in the process of phasing out the SIC System and replacing it with a new system, the North American Industrial Classification System (NAICS).

NAICS included a category explicitly for lasers, but the proposed sub-divisions within that category did not reflect a realistic breakout of industry products. The LEOMA board realized that a Commerce Department survey of domestic and international laser sales could be beneficial, but the proposed survey would not produce useful data. Accordingly, in September 1997, Pat Edsell, Dave Hardwick,[1] and I visited the Census Department in Suitland, Maryland, to discuss the possibility of revising the proposed NAICS laser categories. It was a positive meeting, and the Census Bureau officials agreed to revise the laser categories according to suggestions it would receive from LEOMA.

"It's obvious [the Census Bureau] has not received any input from our industry, and their category definitions have suffered as a result," Hardwick commented at the time. "It's equally obvious that they are very receptive to industry input, if we provide it" (see Figure 6.1).

So LEOMA set about devising subcategories within the laser classification that would make sense. Requests for input went out to all member companies, and several board members—Randy Heyler, John Tracy, and Dean Hodges[2]—volunteered to organize the resulting suggestions into an appro-priate set of subcategories. The fundamental problem was setting the optimum

---

[1]Edsell and Hardwick were executives with Spectra-Physics and Melles Griot, respectively.
[2]Heyler, Tracy, and Hodges were executives with Newport, OptoPower, and Laser Power Optics, respectively.

**FIGURE 6.1**   Dave Hardwick, shown here with OSA executive director Liz Rogan, was LEOMA's president in 1998, and was instrumental in convincing the Commerce Department to adopt LEOMA's suggested laser categories.

level of granularity: too much detail would be cumbersome to collect and analyze; too little would not be useful.

In 1998, Hardwick, Edsell, and I, joined now by Randy Heyler, returned to the Census Bureau with our proposed subcategories (Table 6.1) to the NAICS laser categories. Although the officials we visited appreciated our input, they could not commit to implementing it immediately.

Although we were in regular contact with the officials during the coming months, the proposed subcategories were still not in place by the spring of 1999. One of the largest industry conferences, CLEO, was held in Baltimore in the late spring of that year, and we invited the Census Bureau officials to visit the exhibition held in conjunction with the conference, so they could see for themselves what the lasers that they were classifying looked like. As we wandered the aisles of the CLEO exhibition, we pressed the officials as to when the new categories could be implemented. They told us that implementation was "imminent."

And it was. That summer, the new subcategories in Table 6.1 were officially entered into the NAICS. The Census Bureau planned the first survey to begin in January 2000 of sales in calendar 1999. They would

**TABLE 6.1  Subdivisions of the NAICS "Laser" Category**

| Nondiode Lasers | Diode Lasers |
|---|---|
| $CO_2$ sealed | $\lambda$ shorter than 700 nm |
| $CO_2$ flowing | 700–1000 nm, <10 mW |
| Lamp-pumped solid state, <100 W | 700–1000 nm, 10 mW–3 W |
| Lamp-pumped solid state, >100 W | 700–1000 nm, 3 W–100 W |
| Nondiode-laser-pumped solid state | 700–1000 nm, >100 W |
| Diode-laser-pumped solid state, <100 W | 1001–1700 nm |
| Diode-laser-pumped solid state, >100 W | $\lambda$ longer than 1700 nm |
| HeCd and ion, <1 W | |
| Ion, >1 W | |
| HeNe | |
| Dye | |
| Excimer | |

distribute questionnaires to all U.S. laser manufacturers, who would be required by law to complete them and return them to the Census Bureau. The questionnaire would ask that each manufacturer quantify his domestic and international sales in each of the subcategories of Table 6.1. The Census Bureau would then aggregate the inputs from all companies and release the totals for public consumption.

"This data will be useful to our marketing people in understanding market dynamics," Pat Edsell commented at the time. "And it will help us communicate with the financial community in language they understand."

"It will be a big improvement over existing sales date," agreed Dave Hardwick. "I encourage everybody to respond to the questionnaire as soon as they receive it."

But this story has a disappointing coda. Although the questionnaires were distributed to U.S. laser companies, the data were never collected and analyzed, and results were never disseminated back to industry. LEOMA made several attempts to follow up with the Census Bureau, but was always informed that too few questionnaires had been returned. Although companies were legally required to complete and return the questionnaires, that law was not backed by any enforcement effort. When queried, most LEOMA companies insisted that they had completed and returned the questionnaires. The Census Department was constrained by privacy regulations from commenting on what companies had or had not actually returned the questionnaires. After several attempts to restart this project, the LEOMA board—distracted in 2000 by the association's own financial problems—let the NAICS data fall off the bottom of its priority list. To the best of my knowledge, the NAICS categories in Table 6.1 have never been used to collect meaningful data.

The previous chapter relates the happier story of LEOMA's successful interaction with the federal government to modify export controls on lasers and optics. As mentioned in that chapter, there was a brief period between COCOM and Wassenaar when the United States and other nations considered unilateral export controls. In particular, the so-called Kasich Amendment to the Export Administration Act of 1994 threatened a stifling tightening of the existing controls. Under previous controls, exports to certain countries—countries deemed threatening or hostile to the United States—required Individually Validated Licenses (IVL). The Kasich Amendment would mandate that goods requiring an IVL to any country would require an IVL to all countries. In other words, if you needed an IVL to ship a laser to North Korea, you were also required to obtain an IVL to ship that laser to France. The Kasich Amendment was viewed as extremely onerous by industry, because it was time consuming and expensive to obtain an IVL. And because the controls would apply only to U.S. companies, LEOMA members would suffer an enormous erosion of international competitiveness.

LEOMA sent a delegation including Pat Edsell, Don Scifres, Paul Crosby,[3] and me to Washington, where we met with William Reinsch, the undersecretary of commerce for export affairs, and presented our concerns about the Kasich Amendment. Reinsch assured us that we were not the only ones lobbying against that piece of legislation and, indeed, the amendment eventually failed under a barrage of criticism from U.S. industry. How big a role did our visit with Undersecretary Reinsch play in that outcome? It's hard to say, but it's nice to think it was more than negligible.

I'm afraid other LEOMA efforts with the federal government had pretty close to negligible impact on the eventual outcome of an issue. The research and development tax credit is an example. During the years that LEOMA's executive seminars were held in Washington, industry executives made several visits to congressional offices to lobby in favor of that credit. But LEOMA's voice, compared to many others that were weighing in on the same issue, was surely almost inaudible.

LEOMA's voice was not inaudible in one aspect of the evolution of the national laboratories. The national labs—principally Lawrence Livermore National Laboratory, Los Altos National Laboratory, and Sandia Laboratory—had been known as the weapons labs during the cold war, because their major mission had been to develop nuclear and other weapons for national security. But with the end of the cold war, the pressing need for weapons-related research diminished, and it wasn't clear what should become of the national labs. They employed many of the country's most brilliant scientists and engineers, and there was great reluctance to disband such a valuable national resource. In the early 1990s, the labs' advocates in Congress and

---

[3]At the time, Edsell, Scifres, and Crosby were executives with Spectra-Physics, SDL, and Coherent, respectively.

elsewhere proposed expanding the labs' charter into civilian, nonmilitary technologies. The scientists and engineers who had for decades maintained that nation's lead in military technology could now ensure that the nation remained at the forefront of global civilian technologies.

The labs launched a "National Technology Initiative" (NTI) whose purpose was to facilitate the transfer of the technology developed at the laboratories to commercial companies and universities. The labs, and Livermore in particular, had been very active in the development of new laser technologies, and LEOMA was interested in the NTI and the possibility of commercializing many of the techniques developed at Livermore and Los Alamos.

With the board's approval, I planned to make the NTI the focus of the 1993 LEOMA executive seminar. I registered to attend a Livermore-organized conference in Santa Clara, California, in the summer of 1992 whose purpose was to describe the NTI and to give companies an idea of the technologies that would be available for transfer. The conference was a disappointment. Conference participants indicated that while there was great interest in the labs' technology, the technology exchange mechanism was very cumbersome and awkward, making the actual transfer of technology very difficult. I recommended to the board that our 1993 executive seminar focus on another topic, and the board agreed.

The U.S. laser industry, and the nation's industry in general, viewed the labs' expansion into new civilian fields as something of a double-edged sword. Many of the technologies developed at the labs could have civilian applications, and that could be a blessing if the technologies were transferred to industry for development and commercialization. On the other hand, it could be detrimental if the labs' efforts were parallel to those of industry, essentially placing the labs in competition with private companies.

So LEOMA and its members were interested but wary as the role of the national laboratories evolved. We knew that bills were being introduced in Congress to redefine the role of the national labs by expanding their charter into peacetime activities. Motivated in part by these considerations, LEOMA held is first Washington-based executive seminar in February 1994. Executives attending the seminar scheduled appointments and visited the offices of their senators and representatives. One of the most important purposes of those visits was to emphasize the industry's strong support for expanding the charter of the labs.

LEOMA had prepared a formal position paper on the topic, and our executives left copies of that paper with every office they visited. The paper read in part:

"Traditionally . . . industry has enjoyed a positive and constructive relationship with the national laboratories. . . . However, in the post cold-war era, there is a possibility that the national laboratories will compete with . . . private industry.

LEOMA supports the expansion into non-military technologies. . . . [But] care must be taken to ensure that competition detrimental to the private sector does not occur."

Of particular concern to several LEOMA companies was the giant laser being developed at Livermore for research in inertial-confinement fusion. The concept was eventually to energize that laser with the light from thousands of high-power semiconductor diode lasers. Developing and manufacturing these thousands of lasers was viewed as a great opportunity for the laser industry. The U.S. companies that had commercialized diode lasers during the previous decade had the capability to develop the diode lasers for the Livermore laser. But rather than funding that research at the companies, Livermore absorbed the resources into their own in-house research efforts.

And even as LEOMA executives were visiting the congressional offices, a more troubling situation was coming to light. Livermore was transferring diode lasers to branches of the armed services in an intragovernment arrangement. Private companies that had been supplying the military with diode lasers suddenly discovered that lasers from Livermore were saturating the market.

This situation presented a quandary to laser manufacturers, because at the same time Livermore was usurping part of their market, the lab was also a customer and a potential source for future funding. The companies were alarmed that the lab was directly competing with them—essentially using taxpayer money to drive down the cost of its products in comparison to the companies' products. On the other hand, they were reluctant to complain too loudly, lest they alienate a major customer and a potential source of future research funding.

A way to finesse the situation was presented when Duncan Moore, a professor at the University of Rochester who had recently completed an American Physical Society Congressional Fellowship, arranged for LEOMA to receive an invitation to testify before the House Subcommittee on Energy and Power. That committee was considering a bill to expand the labs' charter into civilian technologies. By testifying on behalf of the entire industry about competition between the national labs and private industry, LEOMA could avoid hard feelings directed at any particular company.

The testimony was scheduled for late March 1994, and during the weeks leading up to that testimony, LEOMA worked very carefully to craft a position statement. Executives from SDL, Opto-Power, and other LEOMA companies worked with me on the wording. We wanted the statement to be supportive of the labs' expansion into peacetime technologies, but not to shy away from condemning any expansion that competed with the private sector. Accordingly, when I gave the testimony on March 24, I opened by emphasizing, "We do believe that a strong and constructive linkage between industry and the DoE labs can enhance the competitiveness of U.S. high-technology companies."

But then I launched into an unambiguous criticism of Livermore for competing with private industry in the market for high-power diode lasers. I explained that the commercial market for these devices began in the early 1980s, and U.S. companies had developed a commanding lead over other nations in the technology. But, I asked rhetorically, "Who do you suppose is the leading manufacturer of these devices in the United States?" There are several private companies in the market, but "By a substantial margin, the nation's leading supplier of these devices is the federally funded Lawrence Livermore Laboratory." In 1994, Livermore was producing high-power diode lasers at rate twice that of the second-largest producer.

Because the manufacturing cost of diode lasers is inversely proportional to volume, U.S. manufacturers were less competitive in the international market, the LEOMA testimony said. Of the 10 U.S. manufacturers in 1990, 6 had either gone out of business or suffered a major downsizing by 1994. And the remaining four had not grown substantially, while diode laser work at Livermore had increased very significantly during the same four years.

I noted that the purpose of the hearings that day was to examine ways of increasing the efficiency of technology transfer from the labs to private companies. "With the greatest of respect, I would like to suggest that the best way to get technology into the private sector is to develop it there in the first place." In the past, government funding for the development of commercial, high-power laser diodes had been diverted from the private sector to Livermore. "Whatever technology does emerge from the laboratories is not superior to what could have been developed in the private sector, had the resources not been diverted," I argued.

I closed by paraphrasing Robert White, the former undersecretary of commerce for technology, who had observed that "bureaucratic expediency driven by the relentless short-term forces of institutional self-preservation has often dominated the choice of activities at the DoE laboratories." Legislation to expand the role of the laboratories into peacetime activities, I urged, should include wording to prohibit the labs from using taxpayer funds to compete directly or indirectly with private industry.

The effect of LEOMA's testimony was very positive from our perspective. Almost immediately, LEOMA representatives began working with the staff of Congresswoman Anna Eshoo, whose district included those LEOMA members in Silicon Valley, to insert wording into the bill that would prohibit the national labs from pursuing technologies that were under development in the private sector. Less than a month after LEOMA's testimony, Eshoo announced that the Power and Energy Subcommittee had approved the revised bill. In her press release announcing the passage, Eshoo explicitly cited the case of diode lasers that had been raised in LEOMA's testimony.

Even more important were the changes at Livermore. The diode laser program there was immediately scaled back, and the transfers to the military of diodes that could be supplied by the private sector were eventually ended. Mike Campbell, the head of Livermore's laser program, requested a letter from LEOMA reiterating our support for the national labs' expansion into peacetime pursuits, provided that those pursuits were not detrimental to companies. LEOMA readily complied with the request.

Then, in September 1994, Campbell and Ralph Jacobs,[4] another leader of Livermore's laser program, visited LEOMA at the SDL offices in Silicon Valley. They described an "NIF[5] Stakeholders' Briefing" being planned at a hotel near Livermore in February. A wide variety of vendors would be invited to the briefing, whose purpose would be to outline NIF's requirements from 1995 to 2002, with the intent to maximize vendors' contributions. It was certainly a dramatic change from Livermore's philosophy in previous years.

In November, several members of LEOMA's government affairs committee, including Don Scifres (see Figure 6.2) and Pat Edsell, joined me in a reciprocal visit to Livermore. Campbell and Jacobs told us that the Stakeholders' Briefing in February would announce a budget in excess of $1.1 billion for the 1995–2002 period, of which 75% would go to outside vendors.

Campbell and Jacobs were also organizing a "NIF Industrial Council" whose purpose would be to monitor the approximately $800 million Livermore planned to spend with outside vendors during the next seven years. Campbell emphasized to us that one of the council's chief functions would be to ensure the Livermore obtained material from outside sources—instead of developing the technology in-house—whenever possible. And Campbell also frankly acknowledged that another purpose of the council was to build NIF's support in Congress.

Campbell offered to visit an upcoming LEOMA board meeting to brief the entire board about Livermore's spending plans. He suggested that we schedule regular meetings between Livermore personnel and LEOMA for updates on the situation. Subsequently, LEOMA appointed George Balogh of Spectra-Physics, and me, to represent LEOMA on the NIF Industrial Council.

---

[4]In earlier years, Jacobs had been active in LEOMA as an employee of Spectra-Physics.
[5]NIF is the National Ignition Facility, a huge laser whose purpose is to achieve "breakeven" and then "ignition." Breakeven is defined as the point where the nuclear energy released is as great as the laser energy applied, and was achieved in February, 2014. Ignition is defined as a self-sustaining fusion reaction, and is not expected for many years.

**FIGURE 6.2** Don Scifres (left) and Jim Hopkins discuss an exhibit at a laser trade show. Scifres was LEOMA's 1996 president, and active for many years in the LEOMA Government Affairs Committee.

Also at the November 1994 meeting at Livermore, Campbell explained that the French inertial confinement program would be required to purchase from European vendors whenever possible, but many of the technologies were unavailable in Europe. He urged LEOMA companies to be aware of the French program, and to bid aggressively on any requisitions it announced.

The NIF Stakeholders' Briefing was held as planned at the Pleasanton (California) Hilton Hotel on February 2 and 3, 1995, and attracted representatives from dozens of companies. They heard numerous Livermore physicists and engineers describe their plans and their equipment requirements for the coming years.

The NIF Industrial Council, whose purpose was to ensure that outside vendors were utilized as much as possible, met on several occasions and was satisfied that Livermore's procurement procedures did, in fact, achieve that goal. I don't know if the lab actually spent the entire $800 million between 1995 and 2002, but the laboratory has become a major consumer of all manner of optical, photonic, and electrical equipment. And LEOMA's 1994 testimony about diode lasers definitely contributed to that outcome.

Domestic sales, both to government agencies and in the private sector, were important to LEOMA members, but in the 1990s nearly half the industry's sales were in the export market. Anything that inhibited those exports was a concern to LEOMA. Export controls, of course, could be a major inhibitor, and many LEOMA resources were directed at them, as discussed in Chapter 5. But tariffs imposed by other countries on U.S. imports were also a factor, and in 1996 LEOMA addressed that issue.

In the mid-1990s, European tariffs on lasers could be as high as 14%, while U.S. tariffs on European lasers were zero. This imbalance was a concern to many LEOMA members, and in March 1996, a LEOMA delegation including Don Scifres, Bernard Couillaud, Pat Edsell,[6] and myself visited the offices of the U.S. Trade Representative (USTR) in Washington to discuss the Information Technology Agreement, which the USTR was currently negotiating with other countries. We explained the existing imbalance in laser tariffs, and urged the USTR officials to include lasers among the items addressed by the ITA. The officials were sympathetic, but vague about any definite commitment.

In January 1997, another LEOMA delegation—Randy Heyler, Chip Greening, Dean Hodges,[7] and I – paid a second call on the USTR offices in Washington. We met with Dorothy Dwoskin, the assistant U.S. trade representative, who told us that an agreement had been reached the previous month in Singapore to cut at least 25% from tariffs on diode lasers by July, and to phase out tariffs altogether by the end of 1999.

How much did LEOMA's visit in March contribute to the December agreement among ITA negotiators? Dwoskin and the other USTR officials we met with were gracious, and implied that the inclusion of diode lasers in the ITA negotiations was a direct result of our visit. I would like to think that was the case, but in reality I do not know. Unlike our experience with the national labs, where the result of our efforts was unambiguous, cause and effect were less clear with the ITA. Perhaps the LEOMA visit did precipitate lower tariffs on diode lasers. Or perhaps diode lasers were included for a completely unrelated issue—perhaps as part of a broad class like "semiconductor devices."

Whatever the reason, we were very pleased that tariffs on diode lasers had been reduced, and LEOMA encouraged its members to write to their respective congressional representatives praising the USTR and Ms Dwoskin, and urging them to encourage the USTR to press for a rapid implementation of the tariff reductions.

---

[6]Scifres, Couillaud, and Edsell were executives with SDL, Coherent, and Spectra-Physics, respectively.

[7]Heyler, Greening, and Hodges were executives at Newport, Omnichrome, and Laser Power Optics, respectively.

By late summer of 1997, 43 nations had signed the ITA, and by the end of the year more than 60 had. But by the end of 1997, attention had turned to the negotiation of an extension of the ITA, or ITA II. The original ITA had included only diode lasers, so tariffs on non-diode lasers were still an issue. In January 1998, Dave Hardwick and I visited the USTR offices once again, this time arguing that *all* lasers have a role in information technology, and therefore all lasers—not just diode lasers—should be included in ITA II. Dwoskin and the other official we met with agreed to that the U.S. proposal to the negotiations would include all lasers.

Unhappily, that part of the U.S. proposal did not survive the international negotiations. Nonetheless, tariffs became a diminishing issue as barriers to trade, in general, fell, and LEOMA was pleased to have played a role in promoting another aspect of international trade.

## THE BENEFIT OF HINDSIGHT

Years after LEOMA ceased to function, years after the last trip of LEOMA executives to Washington, Coherent CEO John Ambroseo told me rather emphatically that he thought those visits of LEOMA executives to the offices of various member of Congress had been a waste of time. He was partially right, and partially wrong. He was right in the sense that many of the issues we addressed were too big for us. By that, I mean the issues—the R&D tax credit, for example—that had powerful forces on both sides of the discussion. When an executive from a medium-size laser company visited his senator's or representative's office to weigh in on the issue, he was received politely, but he had little to no influence on the representative's ultimate decision in the matter.

On the other hand, the trips LEOMA executives made to Washington had significant influence on smaller issues that were nonetheless important to the laser industry. As early as 1990, LEOMA's visits to the Commerce Department helped precipitate the creation of a special license (the "GCT license") that significantly eased the export of lasers to friendly countries. Later, we convinced the Census Bureau to revise its laser categories for the NAICS data, and that could never have been done with letters and phone calls. It isn't clear how much impact LEOMA had on issues like the so-called Kaisch Amendment in 1994, or in the inclusion of diode lasers in the Information Technology Agreement. As explained in Chapter 7, LEOMA may also have contributed to reforms in FDA policy in 1995. What is clear is that LEOMA would have had no influence whatsoever on these issues had its executives not made the journeys to Washington to lobby with the appropriate politicians and officials.

And LEOMA absolutely had a significant impact in altering the strategies of the national laboratories. We would not have been invited to testify before a congressional committee if LEOMA executives had not been in Washington the month before, complaining to their congressional representatives about the labs' current policies.

So, while some of the individual visits to congressional offices and other Washington offices may have been ineffectual, overall LEOMA's occasional trips to the nation's capital did serve a constructive purpose.

# 7

# INTRA-INDUSTRY AFFAIRS

The previous chapters have related the tale of LEOMA's interaction with entities outside the North American laser industry—international standards institutions, the federal government, professional societies, and so forth. But LEOMA also had several projects aimed solely within the industry itself, with the goal of improving the day-to-day operation of the industry. These projects focused on dissemination of information within the industry through several different surveys and seminars, and on reducing the industry's legal bills through an alternative dispute-resolution agreement.

## THE LEOMA EXECUTIVE SEMINAR

As early as January 1989, the LEOMA (then LAA) Board of Trustees was discussing the benefit of some sort of "executive retreat" for members, with the aim providing a valuable opportunity for networking among executives at different companies, and generating information useful to those executives. LEOMA was in the midst of defining its own role within the industry, and a one- or two-day retreat with board members and other industry executives could produce a strong consensus on LEOMA's priorities.

The unification of the European market was a topic of intense interest in 1989 to U.S. companies with European customers, and that subject made

*LEOMA and the U.S. Laser Industry: The Good and Bad Moves for Trade Associations in Emerging High-Tech Industries*, First Edition. C. Breck Hitz.
© 2015 by The Institute of Electrical and Electronics Engineers, Inc. Published 2015 by John Wiley & Sons, Inc.

sense as a focus of any retreat of LEOMA executives. In March, Henry Massenberg[1] agreed to identify several speakers who might address the retreat on that subject. (At that same board meeting, in March 1989, the board decided that "retreat" sounded too much like a boondoggle, and the "LEOMA executive seminar" was born.)

By the middle of 1989, Glenn Sherman[2] had joined Massenberg in finding suitable speakers, and I was directed to prepare a tentative budget for the seminar. LEOMA's treasury was very fragile, and most of the seminar's expenses would have to be paid by the seminar itself, that is, by attendees. Dean Hodges[3] and I prepared a letter that went to 20 companies that were LEOMA members but that were not directly represented on the LEOMA Board of Trustees. The letter inquired about their interest in an off-site, two-day seminar that would offer an opportunity to network with other industry executives, and that would address the issues of "Europe '92." We estimated the fee at $700–1000 including hotel and meals, and predicted the location would be somewhere on the West Coast.

Ten companies responded to the letter; eight indicating a definite interest in the seminar, but two were uninterested. Budget-wise, that was marginal; the per-attendee cost would be prohibitively high if too few non-board companies attended the seminar. The board directed me to find a suitable location and prepare a firm budget for the seminar in the spring of 1990, assuming 20–25 attendees.

After inquiring at several locations up and down the West Coast of the United States, I settled on the Hyatt Regency Hotel in Monterey, California, on Friday and Saturday, March 9 and 10, 1990. The per-attendee cost would be $900, including hotel and meals, and attendees were encouraged to bring spouses, for whom there was no additional fee. And with the help of board members, several speakers had already been identified. Coherent's Hank Gauthier had recruited his neighbor, Ed Zschau, a former congressman and Republican nominee for the U.S. Senate, and in 1989 the president of a Silicon Valley company, to address "The global competitiveness of U.S technology firms." And John Tompkins of Spectra-Physics had asked his vice president for Europe, Klaus Derge, to talk about "The European market for lasers after 1992."

In September 1989, LEOMA sent announcements of the tentative executive seminar to all its members, requesting response by mid-August. We needed at least 24 commitments by that date, or the seminar would be cancelled. Happily, 32 executives committed themselves to participate in the seminar.

---

[1]Massenberg was a Spectra-Physics executive.
[2]Sherman was president of Laser Power Optics.
[3]Hodges was a Newport executive.

I signed the final agreement with the Hyatt Regency, and the first LEOMA executive seminar was set to go.

During the ensuing months, half a dozen additional speakers were lined up, including Jim LeMunyon, the deputy assistant secretary of commerce for export controls, and Samuel Simonsson, the president of the large German firm, Rofin Sinar. A complete list of all speakers at all the LEOMA executive seminars is in Appendix 3.

That first executive seminar was hailed as a resounding success, with no fewer than 37 attendees, 16 of whom were accompanied by their spouses. And although the seminar officially ended at noon Saturday, many attendees stayed another day to enjoy the golf courses and other attractions of the Monterey peninsula.

Another positive outcome of the executive seminar was the announcement by Jim LeMunyon, the deputy assistant secretary of commerce, of the creation of a "GCT license" for export of lasers and other items to COCOM nations. As explained in Chapter 5, COCOM was a group of countries (western Europe, Australia, Japan) that agreed to limit strategic exports to communist countries. Prior to the GCT license, any laser shipped from the United States to another COCOM country required obtaining an individual export license from the Commerce Department, which was an expensive and time-consuming process. After the GCT license was put in place, individual licenses were no longer required to ship to another COCOM country. It was a very significant change, and viewed as a direct result of LEOMA's interaction with the Commerce Department's export control administration.

Finally, at the seminar banquet Friday evening, Glenn Sherman announced that the LEOMA board had unanimously voted to name Milton Chang, whom many considered to be the founder of LEOMA, as an honorary member of the association. Chang, who was present at the banquet, accepted the honor with pleasure.

The seminar generated $13,000 net revenue,[4] and a pair of recommendations for future seminars. Attendees felt more idle time should be scheduled into the seminar to allow for informal conversations and networking. And the seating at meals should be assigned, to facilitate attendees' meeting new people rather than sitting with those they already knew.

Encouraged by the success of the first executive seminar, the LEOMA board moved toward making the seminar an annual event. Planning for a 1991 executive seminar began at the board meeting held during the 1990 seminar. LEOMA's president-elect for 1990, Bob Pressley,[5] volunteered to work with me in identifying speakers and a location for the next seminar. The thinking

[4]Exclusive of LEOMA staff time.
[5]Pressley was president of XMR.

was that the focus would be on business/finance issues affecting laser manufacturers.

The 1990 executive seminar had been exclusively for LEOMA members, but now the question arose, should the 1991 seminar be open to nonmembers? In addition to attracting more paying participants, opening the seminar could familiarize nonmembers with LEOMA and perhaps encourage their joining the association. A broader participation in the seminar would increase its— and LEOMA's—visibility in the industry. The board voted to open the 1991 seminar to nonmembers, and to advertise it aggressively.

The 1991 LEOMA executive seminar was held on March 15–16 in the Mark Hopkins Intercontinental Hotel in San Francisco, and attracted 30 paying attendees. Although the seminar had been broadly marketed as open to employees of nonmember companies, all the attendees were LEOMA members. The speakers, listed in detail in Appendix 1, addressed a number of financial topics of interest to the companies, like the trade-offs between working with venture capitalists and conventional banks.

Two additional speakers addressed topics of global interest to U.S. laser manufacturers. Nancy Mason, the undersecretary of commerce for technology, discussed the Commerce Department's broad efforts to boost the competitiveness of U.S. technology companies. And Jim Davis, a deputy director of Lawrence Livermore National Laboratory, explained how technology transfer from Livermore to U.S. companies would work.

Attendees felt the 1991 executive seminar had been very worthwhile, and by now a broad consensus had formed on the LEOMA board that the seminars should take place every year. But the first two had been held on the West Coast, and board members from non-California companies advocated holding the 1992 seminar on the East Coast. I was directed to identify a suitable location. With the help of Spectra-Physics' Pat Edsell, I settled on the Marriott Hotel in West Palm Beach, Florida. And the 1992 seminar would not be open to companies that were not LEOMA members.

At several meetings in late 1991, the board discussed possible topics for the 1992 seminar. Possibilities included the "Total Quality" concept that was being widely discussed at the time, and an examination of the domestic market for lasers. The first executive seminar, back in 1990, had explored the potential changes in the European market after unification in 1992, but a lot had changed during the ensuing months. The board decided it would be appropriate to have an update on that subject.

Accordingly, several speakers addressed that topic at the 1992 executive seminar in West Palm Beach on March 20–22. Tom Galantowicz, the president of Newport and a LEOMA officer, described the difficulties Newport encountered in the acquisition of a European subsidiary, Micro

Controle. "You wouldn't believe how many helpers you get," he said. Bankers, accountants, attorneys, and others. "Some of them are value added," he said, "And some of them are just added."

Klaus Derge, the Spectra-Physics vice president in charge of the company's European operations, returned to update his discussion of Europe '92 that he had made at the 1990 seminar.

During 1991 a LEOMA-like trade association, the Optoelectronics Industry Development Association (OIDA) had come into existence. Unlike LEOMA, OIDA comprised many of the giant companies—AT&T, IBM, HP—that dominated the U.S optoelectronics market. OIDA's *raison d'être* was to reverse Japan's gain of market share of the global optoelectronics market. It was a mission of great interest to LEOMA members. Arpad Bergh, the OIDA's executive director, spoke at the 1992 LEOMA executive seminar, describing OIDA's goals and their plans for achieving them.

The 1993 executive seminar took place five months after Bill Clinton was elected to his first term as president. There was much buzz about the new, advanced technology policy that would accompany the Clinton administration, and that was a natural focus for the 1993 seminar. The 1992 seminar had been on the East Coast, and the LEOMA board wanted to move it westward again in 1993. I had recently held a meeting of the ISO Laser Standards Committee at the Alexis Park Hotel in Las Vegas, and I recommended that facility. It was comfortable, roomy, and inexpensive.

Shortly after the inauguration, the Clinton administration issued a position paper officially describing the nation's new technology policy. It called for a "strong and sustained support of basic science" and R&D. Since R&D was one of the laser industry's largest markets, LEOMA members were very interested in learning how that would play out. The paper also called for "accelerated investment in advanced manufacturing technologies," which presumably included many laser materials-processing applications.

The 1993 LEOMA executive seminar took place on March 26–27 at the Alexis Park Hotel in Las Vegas. One of the featured speakers was Stephen Cohen of the Berkeley Roundtable on the International Economy (BRIE). BRIE, a small but influential think tank, had been a leading advocate of government involvement in the nation's industrial and technological policy. Many observers expected that the principles espoused by BRIE would become the foundation of U.S technology policy under the Clinton administration. Cohen spoke on "New directions in technology policy under the Clinton administration."

Walter Hoadley of the Hoover Institution and the former chief economist at the Bank of America, discussed likely changes in U.S. economic policy under the Clinton administration. Other speakers addressed issues ranging from importing technology from the former Soviet Union to the North American Free Trade Agreement.

Despite the timeliness of the topics addressed, attendance at the 1993 seminar was down from previous years, with only 22 attendees (and 16 spouses). The $900 fee was in line with previous executive seminars, but in an effort to attract new attendees, a special fee of $550 was offered to attendees from any LEOMA company that had not participated in a previous seminar. That offer attracted no takers.

At the board meeting held during the seminar, I questioned continuing the seminar. With only 22 participants, the 1993 seminar had not paid for itself and had drained resources from the LEOMA treasury. The board members, most of whom had been regular attendees at the seminars, saw a real value and strongly opposed ending them. Dave Farrell,[6] LEOMA's 1993 president, appointed a committee to evaluate the annual executive seminar, and reach a decision on continuing them.

The committee, chaired by Pat Edsell, CEO of Spectra-Physics and including Don Scifres, CEO of SDL and Paul Kenrick, CEO of Melles-Griot, quickly decided that the technology policy of the new Clinton administration was the hot topic of the day, and if there were to be a 1994 executive seminar, it should address that subject and be sited in Washington, DC.

In July, a subset of the committee—Edsell, Kenrick, and I—traveled to Washington to investigate the feasibility of finding suitable speakers and holding a successful seminar there. It was a worthwhile trip. We spoke with many officials in the Clinton administration, in Congress, and at the American Electronics Association. "It is obvious," Edsell wrote after returning, "that having LEOMA's executive seminar in Washington would be very beneficial. We would clearly be able to attract interesting speakers. We would also have the opportunity to visit key officials and express our concerns to them." The committee recommended that the 1994 executive seminar be held in Washington, and include visits by participants to various congressional and administration offices. The board enthusiastically accepted the recommendation.

After visiting several Washington facilities, I settled on the Washington Marriott Hotel as the site for the seminar. The board decided to open the 1994 seminar to employees of nonmember companies (at a significantly higher price: $1500 vs. $975), and the two magazines, *Photonics Spectra* and *Laser Focus World*, agreed to run complimentary ads for the seminar in their November issues.

---

[6]Farrell was an executive at Burleigh.

Now members of Edsell's committee began contacting the people we had visited in July and inviting them to speak at the seminar. As Edsell had predicted, nearly everybody whom we invited accepted the invitation. The speakers included

- Duncan Moore, former director of Institute of Optics at Rochester and the 1994 Congressional Fellow of the American Physical Society,
- Lionel (Skip) Johns, director of Technology Policy at the White House Office of Science and Technology Policy,
- Marc Stanley, associate director of NIST's Advance Technology Program,
- L.N. Durvasula, director of Laser Programs at DARPA,
- Senator Pete Domenici,
- James Jensen, director of Congressional Affairs, Office of Technology Assessment, and
- John Mancini, senior vice president, the American Electronics Association.

An integral part of the agenda of the 1994 seminar would be visits by seminar participants to congressional offices. A month before the seminar, I organized participants into groups according the their congressional districts (a few "groups" consisted of only one person), and asked each group to contact its congressional office to arrange an appointment on the second day of the seminar.

More than three dozen industry executives attended the 1994 executive seminar on February 6–8. Senator Domenici, who had been scheduled to discuss the future of the national laboratories, cancelled his appearance at the last minute, due to pressing congressional issues, but otherwise the seminar came off without a hitch. By its end, there was a strong consensus that the executive seminar should continue on an annual basis, and that it should be located in Washington.

We followed the successful formula we had used to organize the 1994 executive seminar in the 1995 seminar. In September 1994, a LEOMA delegation—Chip Greening,[7] Pat Edsell, and I—visited Washington to invite speakers for the seminar that would take place in the spring of 1995. We were especially pleased that Arati Prabhakar, the director of the National Institute of Standards, was among those who accepted our invitation. NIST was to be instrumental in implementing many aspects of the Clinton technology policy, and its budget had grown from $384 million in 1993 to $935 million in 1995.

---

[7]Greening was president of Omnichrome.

During our two-day visit to Washington, we lined up a full slate of speakers for the 1995 seminar:

- Lee Buchanon, director of the Pentagon's Technology Reinvestment Program, would describe the Defense Department's effort to transfer commercial technology to the private sector.
- William Reinsch, undersecretary of commerce for export controls, would address the evolving nature of export regulations.
- Skip Johns, director of technology policy at the White House Office of Science and Technology Policy, would return to update his presentation in 1994.
- Senator Pete Domenici agreed to take a second try at giving us his perspective on the future of the national laboratories.

On March 12–14, we returned to the Washington Marriott for the 1995 LEOMA executive seminar. Once again, Senator Domenici was unable to make his presentation due to conflicting congressional business, but his aide, Alex Flint, filled in with a well-prepared presentation. Once again, seminar participants spent an afternoon calling on the congressional representatives. LEOMA prepared material dealing with the pending R&D tax credit, and with funding for the National Ignition Facility at Livermore, for the participants to use during their visits.

The 1995 executive seminar was viewed as successful as the one the previous year. To spread the word about the seminar to companies that had not participated, the May 1995 LEOMA newsletter was devoted to a description of the seminar. It included reports, written by participants, of each presentation given at the seminar, and also on each of the congressional visits made by the seminar participants. The information was also provided to the magazines, *Photonics Spectra* and *Laser Focus World*, each of which published a report on the seminar.

During his presentation at the seminar, Skip Johns of the White House Office of Science and Technology Policy, paused to ask the audience, "What are your major concerns?" One that was raised by several attendees was the sluggishness with which the FDA approved new medical devices. Johns asked for further information and specific details, which he could "show the vice president." A week or so after the seminar, his office followed up by asking me for further information. LEOMA provided a handful of examples of slow FDA response. Several weeks later, Vice President Gore's office emerged as a central figure in announced FDA reforms. There were certainly other players involved precipitating those reforms, but I like to think that LEOMA's input played at least a modest role.

At the end of the seminar, LEOMA asked participants to evaluate its value to them an their companies. On the question, "What are the short-term benefits of the seminar to your company?" participants rated the seminar at only 2.7 out of a possible 5; In the short term, they saw limited benefit. But when asked about the long-term benefit, the rating rose to an impressive 4 out of a possible 5. And when asked, "Compared to other things you might do with your time, how valuable is the networking at the executive seminar?" they rated it highly, at 4.4 out of 5.

The upbeat mood continued through the May 1995 board meeting, where there was a consensus that "feeling the pulse" of the federal government was a worthwhile result of the seminar. The board voted unanimously to hold the 1996 executive seminar in Washington.

In September 1995, several board members—Bernard Couillaud,[8] Don Scifres, and Pat Edsell—joined me in what had become an annual trip to Washington to line up speakers for the seminar. It was during this trip that we had a disturbing, if enlightening, visit with a senior staff member on the House Science and Technology Committee. This individual held forth for 40 minutes about his own nonsensical solutions to the problems "Albert Einstein had left on the table." It was disconcerting to me to discover a person seemingly so unfamiliar with mainstream science in a position to influence national science policy.

I was reminded of a comment Alex Flint, the aide to Senator Domenici, had made at the 1995 executive seminar. He said most members Congress did not have a good understanding of technology issues, and weren't interested in acquiring one.

Nonetheless, the Washington trip was rewarded with an interesting slate of speakers for the 1996 seminar. Heading up the program would be Representative Robert Walker (R-PA) who was spearheading the Republican drive in Congress to undo the Clinton administration's technology policy, and Representative Anna Eshoo (D-CA), a fierce supporter of the Clinton policy. Other speakers would address tax policy, and the National Ignition Facility at Livermore.

As discussed later in this chapter, one of the highlights of the 1996 executive seminar was a presentation by John Bates on methods of alternative dispute resolution, especially mediation between companies to avoid litigation. Both Representative Walker and Representative Eshoo were forced to cancel their appearances, but both sent alternates. Substituting for Walker, Doug Comer, chief of staff of the House Committee on Science and Technology, wryly introduced himself as a "spearcatcher" in front of an audience of high-tech executives. He went on to explain that Republicans felt

---

[8]Couillaud was the CEO of Coherent.

that reducing deficit spending was the number-one priority of government, and the Clinton science policy was a major contributor to deficit spending. And in closing, he echoed the comments Alex Flint had made at the previous year's seminar, saying that many members of Congress do not have a good understanding of how science and technology can contribute to the nation's welfare.

Stephanie Vance, an aide to Representative Eshoo, took a dramatically different tack, speaking strongly in support of the Clinton policy and its programs such as Commerce's Advanced Technology Program and the Pentagon's Technology Reinvestment Program. She also advocated a permanent extension of the R&D tax credit. Asked how a small technology industry might influence federal issues of interest to it, Vance offered two suggestions. First, she said, executives should visit Washington and make their concerns and opinions known to their congressional representatives—just as LEOMA was doing. Second, she advised members to invite their Senators and Representatives to visit their companies during congressional recesses. (And LEOMA had arranged a luncheon for Congresswoman Eshoo at Coherent, attended by many LEOMA board members in 1994.)

But attendance at the 1996 seminar fell to only 22, significantly fewer than the previous years. Some of the newness of interacting with the federal government was wearing off. At the board meeting held during the seminar, I suggested giving the Washington location a rest, and holding the 1997 seminar elsewhere—perhaps on the West Coast. That suggestion was thoroughly rejected by the board, whose members felt that the interaction with the federal government was an important LEOMA function.

Bill Clinton won a second term in the November 1996 election, and shortly thereafter Chip Greening, Randy Heyler (see Figure 7.1),[9] Dean Hodges, and I made the annual speaker-recruiting trip to Washington. Because of the election, that trip took place several months later than in previous years, and the executive seminar was likewise postponed. It took place on April 27–29, 1997, in the Washington Doubletree Hotel.

The seminar presentations included a panel three senior staff members[10] from the House and Senate science committees, who gave an informal but intriguing perspective on the haphazard and convoluted process that leads to the creation of federal science policy. The new director of technology policy at

[9]Heyler was an executive at Newport.

[10]Jim Turner, minority staff director of the House Science and Technology Committee, Tom Weimer, majority staff director of the same committee, and Pat Windham, minority staff director of the Senate Science, Technology and Space Committee. Each of the three had more than 20 years' experience in Congress. Although he didn't intend it to be humorous, an off-hand comment by Turner brought down the house. "In the 25 years on the Hill," he said, "I don't believe I've ever seen a long-term solution to anything."

**FIGURE 7.1** Randy Heyler, LEOMA's 1997 president, frequently traveled to Washington, DC to align speakers for LEOMA's executive seminar.

the White House Office of Science and Technology, Henry Kelley, discussed the Clinton administration's goals in science and technology during the second term. And Matt Rhode from the office of the U.S. Trade Representative, briefed the seminar on the pending Information Technology Agreement, which would dramatically reduce European tariffs on U.S.-made diode lasers.

But attendance at the 1997 executive seminar was down to 17 participants. The reality was that resources were being taken from LEOMA's treasury to underwrite an activity that was supposed to pay for itself or even provide an income stream. Nonetheless, the industry executives who attended the seminar thought it very worthwhile and were loath to discontinue it. During a discussion at a subsequent board meeting, the members agreed to launch a coordinated effort to rejuvenate the seminars. Each board member agreed to contact at least two peers at other companies and urge them to participate in future seminars. Pat Edsell announced that Spectra-Physics would start sending mid-level managers to the seminar, and encouraged other companies to do likewise. And the board voted unanimously to continue the executive seminars in 1998.

Following the tried-and-true technique of organizing the seminars, Pat Edsell, Dave Hardwick,[11] and I visited Washington in September 1997, to recruit speakers for the 1998 executive seminar. Among those visited, we called on the office of Representative James Sensenbrenner, the chair of the House Science Committee, where a senior aide, Robert Cook, suggested that we reserve the committee's meeting room on Capitol Hill one afternoon, and invite Sensenbrenner to speak to the group there. We immediately recognized he merit of the suggestion, and followed it.

We took other steps to increase the attractiveness of the 1998 executive seminar. The Latham Hotel, in the heart of Washington's trendy Georgetown neighborhood, was selected as the host hotel. And because participants at previous seminars had consistently requested more time to network with other participants, more free time was scheduled. We deliberately avoided a speaker at the Monday luncheon, and moved the seminar reception from Sunday evening—too early for attendees arriving late—to Monday evening. And after the reception, participants were organized into small groups for dinner at various Washington restaurants.

The COSE Report, a major report by the National Research Council on the state of the art of the nation's photonics technology, is discussed in Chapter 2. The timing was fortunate for LEOMA, because Charles Shank of the Lawrence Berkeley Laboratory, and chairman of the committee that prepared the report, accepted our invitation to describe the report at the seminar. It would be the first public discussion of the report, which had until then been kept under tight wraps. Donald Shapiro, the report's project manager for the National Research Council, also agreed to appear at the seminar. Because the COSE Report was of broad interest to the community at large, LEOMA nonmembers were also invited to participate in the seminar at the same fee as LEOMA members.

The 1998 executive seminar was quite successful, both in terms of attendees and in terms of valuable information disseminated at the seminar. Thirty individuals, including several non-LEOMA members, participated in the seminar. The discussion of the COSE Report was universally perceived as very important. Other speakers, including Duncan Moore of the White House Office of Science and Technology Policy, and Keith Primdahl of the United States [laser isotope] Enrichment Corporation, made valuable contributions.

On the morning of the seminar's second day, participants visited their congressional representatives on Capitol Hill, and then converged on the ornate hearing room of the House Committee on Science and Technology. After lunch, the committee chair, James Sensenbrenner (R-WI), addressed the participants. "Give two speeches a year to [civic groups]," he advised. If

---

[11]Hardwick was an executive with Melles-Griot.

everybody who supported federal funding of science and technology did that, his committee would have no problem obtaining support for a multitude of worthy science and technology projects.

In his talk, Duncan Moore called attention to an issue that would soon become a major project for LEOMA.[12] Manpower shortages, he said, threaten to limit the expansion of the nation's high-technology industries. He presented data showing that America's colleges and universities are producing too few graduates to meet the growing demand of technology industries.

Buoyed by the success of the 1998 seminar, LEOMA enthusiastically moved into planning for a 1999 seminar. Pat Edsell, Dave Hardwick, Randy Heyler, and I traveled to Washington in October 1998, to visit potential speakers. Representative Vernon Ehlers (R-MI), the only member of Congress trained as a physicist, was author of the "National Science Policy Study," commissioned by Speaker Newt Gingrich and expected to play a major role, along with the COSE Report, in the evolution of national science policy. Representative Ehlers agreed to host the 1999 executive seminar in a meeting room of the Rayburn House Office Building, and to describe to us the conclusions of his study.

We also arranged for Representative George Brown (R-CA), the ranking minority member of the House Science Committee to comment on the Ehlers study, as did Representative James Sensenbrenner (R-WI), the committee chair.

As explained in Chapter 5, export controls were being revised in 1999, and LEOMA was concerned that they could again become problematic for the laser industry. Accordingly, we invited William Reinsch, the Commerce undersecretary for Export Administration, to return to the executive seminar in 1999. Representative Chris Cox (R-CA), chair of the Select House Panel on Export Controls, also agreed to address the topic.

Another popular feature of the previous two executive seminars, a panel discussion among senior staff members of the House and Senate science committees, was repeated at the 1999 seminar. These unscripted and informal discussions allowed seminar participants an "insider's view" of the politics that affected the development of national science and technology policy. Breaking participants up into small groups for dinner at various Washington restaurants, also popular at the 1998 seminar, was repeated in 1999.

Despite the careful planning and the influential speakers at the 1999 executive seminar, it drew only 19 participants, and again was a drain on the LEOMA treasury. At its September 1999 board meeting, LEOMA agonized over whether or not to continue organizing the seminars. As described in Chapter 1, the LEOMA treasury was already quite fragile,

---

[12]LEOMA's efforts to boost education in science and technology are discussed in Chapter 4.

without the additional losses of the executive seminar. Chapter 2 describes the Coalition for Photonics and Optics (CPO) and the fact that I was to chair it in 2000. I knew I wanted to redirect CPO's attention to educational issues, so I suggested to the LEOMA board the possibility of a seminar in 2000 focusing on manpower issues, and sponsored jointly by LEOMA and CPO. The board directed me to pursue that possibility. If the CPO did not want to participate, then the board decided LEOMA would skip the executive seminar in 2000, and make it a biannual event after that.

But the CPO did like the idea of a joint seminar focusing on manpower shortages in science and technology. So I worked with staff members of the professional societies comprising the CPO in organizing a program examining the pervasive problem of looming manpower shortages and potential solutions to that problem. These solutions ranged from federal programs to boost enrolment in science and technology programs to things that individuals could do with their local schools to increase interest in these fields.

The seminar took place in June 2000 at the Washington Hyatt Hotel. The agenda included several speakers from high-tech companies describing their approaches to the problem, including LEOMA's Bob Phillippy, who described the PowerPoint presentation he had developed for high school and college audiences.[13] Other speakers from the Education Department and from the National Science Foundation described those institutions programs addressing the issue. And Duncan Moore, who had become one of the nation's champions of high-tech education, emphasized the need to find solutions if the nation was to maintain its competitiveness in the international marketplace.

The seminar was successful and well attended, with three dozen participants. But it was disappointing to me that, despite the urgency of the topic to LEOMA companies, only a handful of those attendees were from LEOMA.

And when the LEOMA board considered undertaking a 2001 executive seminar, there really was no choice. The LEOMA treasury could not sustain the loss associated with a seminar attracting fewer than two dozen attendees, and it was very unlikely that any seminar LEOMA could organize—no matter how valuable the seminar was considered to be—would break even financially. The year-2000 executive seminar, held jointly with CPO, was LEOMA's last executive seminar.

## THE LEOMA ECONOMIC SURVEY

At one of its earliest meetings, in 1988, the LEOMA (then LAA) board discussed the possibility of collecting market data from members and

---

[13]Phillippy's PowerPoint presentation is described in Chapter 4.

compiling a report that would accurately reflect real-time market conditions. The only market information that was available at the time was an annual survey published by *Laser Focus World* magazine, and it was widely understood to include a good measure of Kentucky windage.

If real data could be collected in full confidence from the association's members and compiled into a report, that report would be far more authoritative than the *Laser Focus* report. At the board's instruction, I contacted several accounting and survey firms and inquired about their costs to perform such a study. The answers that came back ranged from $10,000 to more than $50,000, a price that greatly exceeded the resources in the LAA treasury. The board voted to drop the project.

Several years later, during a visit to the American Electronics Association, we learned that a survey of the members' sales was one of the most appreciated services that organization offered its members. At its November 1991 meeting, the LEOMA Board took that fact to heart and unanimously voted to undertake a market survey. Rather than hiring an outside firm to conduct the survey, the board decided that LEOMA would hire an accounting firm to collect the members' data in the strictest confidence, and forward only the totals—no company-specific data—to me. I was to generate the report that would be distributed to members.

After contacting several firms, I settled on Deloitte-Touche to collect the data. A committee of board members, including Bob Gelber,[14] Pat Edsell, and Dave Farrell, set about designing a questionnaire for Deloitte to distribute to members.

Both market data and economic data would be collected. Market data would include orders and shipments broken into three broad market segments: scientific, OEM, and end user. These data would be further broken out by geographic region, but there would be no finer granularity. In other words, an optical bench sold to a university in Chicago would be in the same category as a laser sold to that school. More detailed data would have been useful, but the companies were unwilling to share their own data at a greater level of detail.

The economic data would provide a snapshot of the industry's economic health. Parameters such as total sales, gross margin, pretax income, cost of goods sold, and number of employees would be measured.

The questionnaire for the first LEOMA Economic Survey was distributed to members in early 1992, and 20 companies returned their completed questionnaires to Deloitte. The results, showing data for calendar years 1990 and 1991, were distributed at the May board meeting. But before handing out the report, LEOMA's 1992 president, Bob Gelber, passed around a "quiz" for the board members. The quiz asked the directors what they

---

[14]Gelber was vice president at Coherent.

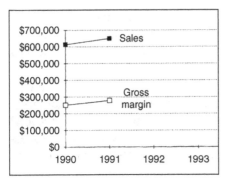

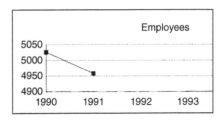

**FIGURE 7.2**   An example of the economic data collected in the first LEOMA Economic Survey.

thought the numbers in the report would be. "If we just hand out the report, everybody will say, '*Yeah, I knew that*,'" Gelber explained. "This way, you'll really know if the survey gives you new information."

In general, the directors displayed a good qualitative understanding of the industry and its markets, but surprisingly little quantitative precision. In fact, the estimates of various economic parameters varied by factors of two and three from one person to the next . . . and from the reality reflected in the survey. Nearly all the directors expressed surprise at the divergence between their guesses and the accurate numbers in the survey (see Figure 7.2), and all agreed that the survey would add important new information to their understanding of electro-optics markets and finances (see Figure 7.3).

There was an interest in expanding the survey from North American markets to worldwide markets for lasers and electro-optics, and to this end LEOMA contacted trade organizations in Japan, Germany, and the United Kingdom. These organizations indicated a lack of interest in participating in the LEOMA survey, but the board decided that questionnaires for the 1993 survey should be sent to every company listed in the *Laser Focus Buyers' Guide*—essentially, every significant electro-optics manufacturer in the world. I proposed that the 1993 survey should be distributed without charge to all companies that submitted data for the survey, but the board voted that nonmembers would pay $500. Nonmembers that did not submit data could purchase the report for $5000.

Unfortunately, all this effort to expand the coverage of the LEOMA Economic Survey came to naught. No nonmember companies ever participated in the survey, so the information reflected only the economic data of LEOMA member companies. Still, those companies represented well over three-quarters of the North American laser market, so the data were not insignificant.

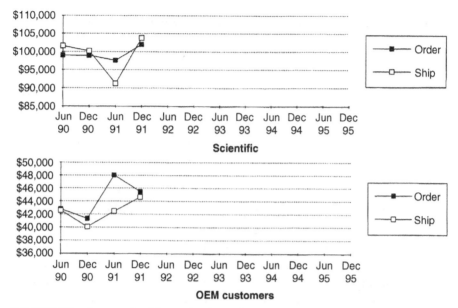

**FIGURE 7.3**    An example of the market data collected in the first LEOMA Economic Survey.

Many directors wanted the survey to divide sales into finer categories, perhaps as fine as breaking out sales by types of laser. But others on the board were unwilling to share their sales data to that level of detail. A compromise was reached for the 1993 survey: sales would be broken out into three categories, lasers, optics, and other "electro-optics related" equipment. Now an optical bench sold to a school in Chicago would be distinguished from a laser sold to that school.

The drawback of this change in scope was that the data for 1990 and 1991 would have to be collected again, showing these new categories. The companies agreed to provide this new data, but the board emphasized that there would be no changes in scope after 1993. In the future, the data for 1990 through 1992 would be compared with data for the current years.

Even though the scope of the survey was locked after 1993, there were still complications as new companies joined the survey, or companies that had participated previously dropped out. When a new company joined the survey, it was required to provide data for all the years covered by the survey, which would eventually become five years. This requirement was a significant deterrent to new companies' joining. To compensate, the board decided that a company that provided less than five years' data to the survey would receive a copy of the survey, but its data would not be included until it could provide data for five years.

On the other hand, when a company dropped out of the survey, all its data for previous years had to be withdrawn from the results for those years. Altogether, the economic survey was a project that absorbed a considerable portion of LEOMA's resources.

And providing the data was a chore at the companies. None of them kept their own records in categories that could be easily converted to the categories in the LEOMA survey, so it became a time-consuming task to complete the LEOMA questionnaire. By 1995, I had to remind many companies multiple times to return their questionnaires. On several occasions, the report was delayed because crucial companies had failed to meet the deadline for submitting data.

At a 1996 board meeting, I formally proposed ending the economic survey. It had become so difficult to extract data from the companies in a timely manner that I questioned the survey's value to those companies. But several companies—led by Newport and Melles-Griot, according to the meeting minutes—strongly opposed discontinuing the survey. The results were very valuable, they said, and were carefully considered as the companies planned their market strategies.

But input for the survey continued to be sluggish. The board would set a deadline, and as that date approached, I would learn from Deloitte which companies had not submitted data. And then I would call the tardy companies and nag. This process would be repeated over and over, until the last company had finally submitted its data.

Abandoning the survey was discussed at several board meetings between 1997 and 1999, but the directors voted every time to continue. The board agreed that each director would personally be responsible for ensuring that the data from his company was submitted by the deadline. But the directors, who were in most cases the CEOs of their companies, often had more pressing matters to attend to, and timely data collection remained a sticking point.

At the board meeting in May 2000, I had to announce that the economic survey for that year was delayed because four major companies had not submitted their data from 1999. I suggested extracting those companies' data from the data for the previous four years, and publishing the 2000 report with the remaining data. But the four companies represented a major slice of the marketplace, and the board directed me to renew my efforts to obtain data from those four companies.

By the October board meeting, several major companies still had not provided 1999 data for the survey. I argued that if the survey really were that valuable to the industry, it would not be so difficult to obtain raw data from the participating companies. I urged the board to discontinue the survey, and the board, in a non unanimous vote, agreed.

## THE LEOMA COMPENSATION SURVEY

I had begun teaching my course, *Understanding Laser Technology*,[15] at LEOMA companies in 1994, and in that process I had become acquainted with the human resource managers at many of those companies. In my casual conversations with these managers, they often complained about the lack of good, comparative data on compensation levels for employees specializing in lasers and optics. There were plenty of data about the compensation levels of electrical engineers, for example, or customer service technicians. But laser engineers and precision-optical technicians had distinct skill sets, and the HR managers wanted hard information about their compensation levels across the industry.

In 1996, LEOMA set about designing a compensation survey for laser and optics engineers and technicians. We created two committees, one headed by Ken Johnson of Spectra-Physics, and made up of HR managers from several other laser companies, to design the survey for laser engineers and technicians. The second committee, composed of HR managers from half a dozen optics companies and headed by Vicki Hoffman of Coherent, designed the survey for optics engineers and technicians.

These two committees first came up with job titles that included the employees for whom they wanted data, and then devised rigorous definitions of the work performed by employees with these titles (see Figure 7.4). The rigorousness was necessary to ensure that the data were consistent from one company to the next.

The LEOMA board endorsed the concept of compensation surveys, but instructed me that the surveys would have to pay for themselves. Fees for

---

**COMPBASE®**
1997 LEOMA – *LASER & OPTICS INDUSTRY COMPENSATION SURVEY*

WESTERN
MANAGEMENT
GROUP

---

**2130 LASER SYSTEM FINAL TEST TECHNICIAN 3** - *Senior Level*

Performs a variety of technical functions related to the calibration, alignment, fabrication, assembly, modification and final test of laser assemblies and optics. Determines and may develop test specifications, methods and procedures from blueprints, drawings and diagrams. May complete rework on assemblies and/or systems as a result of testing. May also prepare technical reports summarizing findings and recommending manufacturing process or design changes. Works from general concepts and design parameters and develops test procedures and documentation to meet production requirements. Performs complex analysis and non-standard tests where considerable judgment and initiative are required in resolving problems and making recommendations. Uses a wide variety of specialized test equipment and computer programs to obtain desired results. Will typically train and provide work leadership to lower level employees. May occasionally visit customer sites to install or trouble shoot new systems.

**Experience and Training:** AA degree or equivalent vocational training with five or more years or directly related experience required.

**FIGURE 7.4**    A job description for one of the 74 job titles in the 1997 LEOMA Compensation Survey.

---

[15]The course is described in Chapter 4.

participation in the surveys would have to cover all costs, and no resources from the LEOMA treasury could go toward the survey. My tentative budget for the projects indicated that the fee would have to be somewhere around $600 to $900 for each survey, for each participating company. My feeling was that a fee that high would prevent many companies from participating. With this in mind, I approached the HR managers of several of the larger LEOMA companies—individuals who enthusiastically supported the surveys and who had been instrumental in their creation—and asked them for voluntary financial support. A handful of companies complied with my request and kicked in a thousand dollars apiece. With this underwriting, I was able to lower the per-company, per-survey fee to $350.

LEOMA hired a professional compensation consulting firm, Western Management of Los Gatos, California, to collect the data in strict confidence and analyze the results. During the autumn of 1996, LEOMA distributed the questionnaires that had been designed by Johnson's and Hoffman's committees, with instructions to return the completed questionnaires to Western Management. No fewer than 36 companies did so, and Western Management published its analysis in December 1996 (see Figure 7.5).

"I think it's very useful," said Lisa Borzachillo, HR director at Melles-Griot, reflecting the opinions of HR managers at other companies who received the report. There was strong agreement that the surveys should be repeated in 1997.

| Total Companies Matching Job 13 | ESTABLISHED RANGES | | | ANNUAL CURRENT PAID RATES | | | |
|---|---|---|---|---|---|---|---|
| | MINIMUM BASE | CONTROL POINT | MAXIMUM BASE | BASE PAY | CASH #1 | CASH #2 | TOTAL CASH |
| A Weighted Average: | $33,605 | $41,075 | $50,700 | $39,443 | $1,535 | $3,130 | $42,206 |
| B Simple/Unweighted Average: | 30,457 | 36,829 | 44,370 | 39,537 | 2,326 | 7,011 | 43,848 |
| C Lowest: | 23,608 | 26,229 | 28,850 | 24,461 | 687 | 1,399 | 30,529 |
| 10th Percentile: | 23,608 | 26,229 | 28,850 | 32,240 | 687 | 1,622 | 33,207 |
| 25th Percentile: | 25,794 | 32,500 | 38,500 | 36,525 | 889 | 1,684 | 38,916 |
| D 50th Percentile/Median: | 30,695 | 36,685 | 44,113 | 39,395 | 952 | 2,857 | 41,880 |
| 75th Percentile: | 34,055 | 42,106 | 52,633 | 42,369 | 967 | 3,118 | 44,697 |
| 90th Percentile: | 36,061 | 42,425 | 52,699 | 45,000 | 967 | 3,436 | 47,067 |
| E Highest: | 36,244 | 44,472 | 52,785 | 65,000 | 5,349 | 31,543 | 101,892 |
| F Number Companies Reporting: | 8 | 8 | 8 | 13 | 3 | 7 | 13 |
| G Number of Employees: | 64 | 64 | 64 | 73 | 7 | 61 | 73 |
| H Earnings Mix (Only For Those Receiving Other Cash): | | | | 92.9% + | 0.4% + | 6.8% = | 100.0% |
| I Percentage of Total Employees Eligible: | | | | | 11.0% | 84.9% | |
| J Percentage of Companies With Eligible Employees: | | | | | 30.8% | 61.5% | |
| K Percentage of Eligible Employees Who Receive Payment: | | | | | 87.5% | 98.4% | |
| L Percentage of Total Employees Receiving Additional Cash: | | | | | 9.6% | 83.6% | |
| M Other Cash/Base Pay - All Employees: | | | | | 0.4% | 6.6% | |
| N Other Cash/Base Pay - Only Employees Receiving: | | | | | 4.2% | 7.8% | |
| O Salary Range Spread: | Weighted Average: | 51.0% | Simple Average: | 45.5% | | | |
| P FLSA Treatment: | Exempt: | 4.1% | Nonexempt: | 95.9% | | | |
| Q Job Match: | Less Complex (-): | 30.1% | Very Close Match (=): | 68.5% | More Complex (+): | 1.4% | |

**FIGURE 7.5** Compensation data for the job title described in Figure 7.5.

But, learning from the results of the 1996 survey, Johnson and Hoffman and their committees instituted several improvements in the 1997 survey. At the suggestion of Western Management, the two distinct reports in 1996—one for optics and one for lasers—were combined into a single report. And because the larger companies were still underwriting the project, the cost for the combined report would remain at $350.

Another improvement in 1997 was the inclusion of new job titles. "We're asking everybody who took part in the 1996 survey to provide us with descriptions of jobs at their companies that didn't fit well into the survey," Hoffman announced. Altogether, the questionnaires for the 1997 compensation survey listed 74 different job titles for electro-optics engineers and technicians, and each one was unique to the laser/optics industry.

Wishing to gather as much data as possible, LEOMA opened the 1997 survey to all North American companies, not just LEOMA members. We invited the American Precision Optics Manufacturers' Association (APOMA) and the Corporate Associates of the Optical Society of America (OSA) to advise their members of the survey's availability, and many members of those organizations did, in fact, participate in the 1997 survey. Altogether, 34 companies participated in the 1997 survey, slightly fewer than in the previous year—but that number is misleading. Of the 34 in the 1997 study, 15 had not participated in 1996.[16] Most of these 15 were large, established companies with many employees, while the dropouts were mostly small optics shops with only several employees.

The compensation survey continued successfully through 1998 and 1999, experiencing little of the difficulty in collecting data that hampered the economic survey. But by 2000 the industry was starting to suffer from a market slowdown, and hiring employees became a less-feverish undertaking. With less hiring, there was less need for compensation data. The 2000 LEOMA Compensation Survey was published on schedule in December of that year, but as 2001 began, it became clear that the large companies were no longer interested in the expense of underwriting the survey. Without that underwriting, the survey could not continue.

## THE LEOMA HUMAN RELATIONS SEMINAR

In 1996, as the HR professionals at various laser and photonics companies worked together creating and improving the LEOMA Compensation Survey,

---

[16]Unlike the economic survey, the compensation survey did not attempt to show year-to-year trends. Each compensation survey presented data for only one year, so the complex adjustments mandated when a company joined or left the economic survey were not necessary in the compensation survey.

they realized that they had a lot of information it would be useful to share with one another. Several of them—including Ken Johnson and Spectra-Physics, Vicki Hoffman at Coherent, and Ingrid Stern at Newport—offered to help LEOMA organize a seminar for HR managers, similar to the LEOMA executive seminar for the companies' top executives.

The LEOMA board was supportive of the concept, and encouraged me to follow through with the HR managers. The first LEOMA Human Relations Seminar took place on February 5–6, 1997, at the Stanford Park Hotel in Menlo Park, California. It was cosponsored by the Optical Society, and open to the OSA's corporate associates as well as LEOMA members. (The corporate associates' fee, at $850, was higher than the $650 fee for LEOMA members.)

The agenda included several outside speakers, for example, Tom Bassett from a San Francisco accounting firm, Pension Metrix, discussing "401(k) plans for small businesses," and Leno Pedrotti of the Center for Occupational Research and Development, discussing "The CORD Skill Set for Laser and Optics Technicians." But most of the program was devoted to talks by, and panel discussions among, the HR managers themselves. Formal topics on the agenda included "variable pay programs," "In-house Training Programs," and "Helping your company meet its strategic goals."

Several interesting subjects emerged during the Q&A sessions that followed each presentation and panel discussion. It turned out that there was a strong interest in the companies' car policies: under what conditions and with what controls did the different companies provide their top executives, and their field-service engineers, with automobiles? And relocation expenses: how were new employees reimbursed for their moving expenses? There were great differences in these issues from one company to the next, but all the HR managers appreciated a window into other companies' policies.

So there was no question about repeating the HR seminar in 1998. It was scheduled at the Sheraton Hotel in La Jolla, California, on April 6–7, 1998. As had been the case at the first seminar, most of the time was devoted to formal and informal discussions among the attendees. One exception was the presentation by John Bates, the same attorney who had addressed the 1996 LEOMA executive seminar on the topic of alternative dispute resolution. Bates described to the HR manages how ADR techniques can be successfully applied in employee-employer disputes.

Newport's Ingrid Stern described her company's implementation of computerized HR information systems, a technique that was coming into vogue that year. And there was an intense interest in training issues. Both Ken Johnson (Spectra-Physics) and Vicki Hoffman (Coherent) discussed their companies' approaches, and I reviewed the LEOMA Laser Short Course.

As they had the year before, the HR executives attending the 1998 seminar were enthusiastic about organizing a similar seminar in 1999. The 1999 seminar was sited in San Diego's Town and Country Hotel, and the format was changed from previous years. Instead of presentations by the HR managers at large LEOMA companies, the 1999 seminar featured panel discussions among managers from both large and small companies. Topics ranged from executive compensation and technical training issues to recruiting and retention.

In 1999, the U.S. laser/photonics industry was enjoying what would turn out to be the peak of a boom-and-bust cycle. Companies were desperate to hire capable employees, so it made sense for the HR managers to focus the 2000 seminar on the issues of training and employees. As discussed several paragraphs above, the 2000 LEOMA executive seminar was jointly sponsored by CPO and LEOMA, and focused on manpower issues in general. The 2000 LEOMA HR Seminar was folded into this event. As explained above, that seminar was successful, with more than three dozen attendees, but surprisingly few from LEOMA companies.

But by 2001, the boom had largely turned to bust for the U.S. photonics industry. The LEOMA board instructed me to drastically reduce the amount of my time devoted to the HR seminar. Several of the most active HR executives at large companies tried to organize a seminar in September 2001, and LEOMA sent out invitations for it, but the response was negligible, and the seminar was cancelled.

## THE ALTERNATIVE DISPUTE—RESOLUTION AGREEMENT

There was an enormous amount of patent litigation among laser companies in the 1980s and 1990s. Probably the most expensive example was a bare-fisted squabble between Spectra-Physics and Coherent involving ion lasers, but there were many others. It's difficult to quantify how much money was transferred from the laser industry to the legal industry during those years, but knowledgeable estimates start at a million dollars a year and range upward.

I was lamenting this sad situation to my father-in-law, an attorney, during a casual dinner party conversation one evening in 1995. He offered to put me in touch with a colleague of his, John Bates, an attorney with JAMS/Endispute, one of the nation's leading practitioners of alternative dispute resolution. Bates explained to me that the primary alternative dispute-resolution (ADR) techniques are mediation and arbitration. Mediation, he said, was an informal negotiation between disagreeing parties, facilitated by a neutral third party. Arbitration was a more formal process, during which the parties present their cases to a neutral arbitrator, who then issues a binding, final resolution. ADR costs are typically one-quarter to one-half those of formal litigation, and ADR techniques can usually settle disagreements in months, as opposed to the years often required for litigation. Bates felt that an industry-wide ADR Agreement would be a boon to the laser industry, and he was enthusiastic about helping LEOMA design such an agreement.

LEOMA's executive committee was equally enthusiastic about the concept when I proposed it to them in the fall of 1995. They authorized me to invite Bates to the December board meeting to outline the approach he would recommend.

Bates's presentation was the main topic of the December board meeting. He explained the advantages of ADR, and Scott Miller, Coherent's in-house counsel, backed him up. Miller said he was familiar with only one occasion when Coherent had used ADR, but it had settled an "enormously complex" case in only a few days, and "saved us a huge amount of money."

The board voted unanimously to continue the development of an industry-wide ADR Agreement for the laser industry. There were a couple questions about what the agreement should specify. The board decided that the ADR Agreement should cover all potential litigation among LEOMA companies, not just patent issues. Then the board discussed exactly how the ADR process should work. Many companies were uncomfortable with the concept of binding arbitration, but were comfortable with mediation. Bates noted that, in his experience, mediation alone had resolved probably 90% of the disputes in which he had been involved. So the agreement would specify that parties in disagreement would first sit down and attempt to negotiate an agreement by themselves. If they were unsuccessful, they would take the case into mediation. If mediation failed, then the parties would be free to initiate litigation.

Bates drew up a proposed agreement, and LEOMA circulated it to the membership. The response was favorable, but to better acquaint the LEOMA companies with the concept of mediation, Bates attended the 1996 LEOMA executive seminar and staged a mock mediation. Seminar attendees were assigned roles as adversaries and their attorneys. Both sides presented their cases in front of the audience, then Bates sent each side out to the hallway while he reasoned with the other. In the end, Bates produced a settlement that neither side loved, but both termed it "acceptable."

"The benefit for me was being able to see how the mediation process works for both sides," commented Bob Gelber. "I would be much more comfortable entering into a mediation now than I would have been before."

Two additional issues arose as the proposed agreement circulated among LEOMA members during the spring of 1996. The first arose because the document did not specify whether the agreement would also bind subsidiaries to submit to negotiation and mediation prior to initiating litigation. At its July meeting, the LEOMA board agreed to insert wording to ensure that the agreement would apply to a signer's subsidiaries.

The second issue was more subtle. Coherent's attorney, Scott Miller, pointed out that the proposed agreement was silent about what happens in the event that mediation is unsuccessful in resolving a dispute. Conceivably, the initiating party could lose the advantage of naming the jurisdiction for litigation, if the responding party to the mediation were to file the first formal lawsuit. Miller suggested that this issue could be fixed by giving the party that initiates mediation the exclusive right to file suit in the first 30 days after the mediation ends. The board agreed to this solution, and the appropriate wording was inserted into the agreement.

By mid-1996, nearly all LEOMA members, including all companies with sales over $4 million, had signed the LEOMA Alternative Dispute-Resolution Agreement. That document is included in Appendix 4.

Did the LEOMA ADR Agreement ever get used? Companies don't like to talk about legal problems, so the agreement may have come into play unbeknownst to me. I am aware of one case where mediation was invoked. A larger LEOMA company acquired a smaller one, and there was a dispute about details of the acquisition. The dispute was settled under mediation.

But how frequently the ADR Agreement was formally invoked may be beside the point. Its mere existence no doubt encouraged companies to negotiate among themselves to settle disputes, rather than proceed immediately to legal hostilities. To the best of my knowledge, there was no litigation between LEOMA members from the time the agreement went into effect, until LEOMA ceased its operations.

## THE BENEFIT OF HINDSIGHT

It took several years to get the executive seminar situated in Washington, DC which probably would have been a better location from the beginning. And it took us several years in Washington to figure out that meeting space was available on Capitol Hill. Meeting space in hotels is expensive, while space on the Hill is usually free. But a more important advantage is that speakers—members of Congress and their aides—are much more likely to show up if they can reach the location with a simple ride on the Capitol subway, rather than a cross-town excursion in an automobile.

The American Electronics Association told us one of the most valued services they provided members was their survey of members' markets. LEOMA's Economic Survey turned out to be less appreciated by its members. A serious hindrance was the difficulty in collecting data from member companies. In most cases, it was the marketing and strategic planning people in a company who made use of the survey data. But responsibility for transforming the company's raw data into a format suitable for the survey, and submitting those data to Deloitte, fell to the company's accounting staff. So the people who had to do the work to make the survey successful were not the people who benefited from a successful survey.

The compensation survey and the series of HR seminars were useful when companies' HR departments were vital to their growth during a boom cycle. But when the boom ended, the need for these tools diminished drastically.

Alternative dispute-resolution agreements are useful instruments within an industry, and one can only speculate about the money that might have been saved if the laser industry had implemented one a decade earlier. A year after the LEOMA ADR Agreement had been implemented, in July 1997, the *Wall Street Journal* reported that many of the nation's largest companies— PepsiCo, Kraft Foods, and Kellogg Co. in the food industry, and DuPont and W.R. Grace in chemicals, for example—had signed pacts agreeing to mediate disputes with industry rivals. The article went on to proclaim, "The movement promises to alter the course of corporate litigation and significantly cut legal costs."

# APPENDIX 1

# LEOMA OFFICERS

*LEOMA and the U.S. Laser Industry: The Good and Bad Moves for Trade Associations in Emerging High-Tech Industries*, First Edition. C. Breck Hitz.
© 2015 by The Institute of Electrical and Electronics Engineers, Inc. Published 2015 by John Wiley & Sons, Inc.

| | 1987 | 1988 | 1989 | 1990 | 1991 | 1992 | 1993 | 1994 | 1995 |
|---|---|---|---|---|---|---|---|---|---|
| **President** | Len DeBenedictis | Jon Tompkins | Dean Hodges | Dale Crane | Bob Pressley | Bob Gelber | Dave Farrell | Pat Edsell | Des Bradley |
| **President-elect** | | Dean Hodges | Dale Crane | Bob Pressley | Bob Gelber | Dave Farrell | Pat Edsell | Des Bradley | Don Scifres |
| **Past president** | Len DeBenedictis | Jon Tompkins | Dean Hodges | Dale Crane | Bob Pressley | Bob Gelber | Dave Farrell | Pat Edsell | |
| **Secretary** | | Dale Crane | Bob Pressley | Bob Gelber | Dave Farrell | Tom Galantowicz | Mark Fukuhara | Don Scifres | Randy Heyler |
| **Treasurer** | Bill Kern | Bill Kern | Bob Gelber | Henry Massenberg | Tom Galantowicz | Pat Edsell | Des Bradley | Randy Heyler | Chip Greening |
| | | Δ First OPTCON | Δ LAA/ LEOMA | | | | | | |

| 1996 | 1997 | 1998 | 1999 | 2000 | 2001 | 2002 | 2003 | 2004 | 2005 |
|---|---|---|---|---|---|---|---|---|---|
| Don | Randy | Dave | George | Bob | Steve | Steve | John | | |
| Scifres | Heyler | Hardwick | Balogh | Phillippy | Sheng | Sheng | Ambroseo | | |
| Randy | Chip | George | Bob | Steve | Dana | John | | | |
| Heyler | Greening | Balogh | Phillippy | Sheng | Marshall | Ambroseo | | | |
| Des | Don | Randy | Dave | (None) | Bob | Steve | Steve | John | |
| Bradley | Scifres | Heyler | Hardwick | | Phillippy | Sheng | Sheng | Ambroseo | |
| | Dave | Thom | Len | | (None) | Brian | | | |
| | Hardwick | Large | Marabella | (None) | | Lula | | | |
| Bernard | Thom | Len | Tom | (None) | Lee | Lynn | | | |
| Couillaud | Large | Marabella | Mahony | | | Strickland | | | |

# APPENDIX 2

# ISO LASER STANDARDS

**Topics Covered by Standards Created in the ISO Committee on Laser Standards (ISO TC172 SC9)** In most cases, the topics listed here are addressed in two or more actual standards. For example, the topic of "Test methods for laser-induced damage threshold" includes separate standards addressing (1) definitions, (2) threshold determination, (3) assurance of power-handling capabilities, and (4) inspection, detection, and measurement.

| | |
|---|---|
| Lasers and laser-related equipment | Vocabulary and symbols |
| Lasers and laser-related equipment | Test methods for laser beam widths, divergence angles, and beam propagation ratios |
| Lasers and laser-related equipment | Standard optical components |
| Lasers and laser-related equipment | Laser device — Minimum requirements for documentation |
| Lasers and laser-related equipment | Test method for absorptance of optical laser components |
| Safety of machinery | Safety and noise requirements of laser processing machines |
| Lasers and laser-related equipment | Test methods for laser beam power, energy, and temporal characteristics |
| Lasers and laser-related equipment | Test methods for laser beam parameters — Beam positional stability |
| Integrated optics | Vocabulary |

*LEOMA and the U.S. Laser Industry: The Good and Bad Moves for Trade Associations in Emerging High-Tech Industries*, First Edition. C. Breck Hitz.
© 2015 by The Institute of Electrical and Electronics Engineers, Inc. Published 2015 by John Wiley & Sons, Inc.

| | |
|---|---|
| Lasers and laser-related equipment | Test method and classification for the laser resistance of surgical drapes and/or patient protective covers |
| Lasers and laser-related equipment | Determination of laser resistance of tracheal tubes |
| Lasers and laser-related equipment | Test methods for laser beam parameters — Polarization |
| Lasers and laser-related equipment | Test methods for laser beam power (energy) density distribution |
| Lasers and laser-related equipment | Test methods for the spectral characteristics of lasers |
| Optics and optical instruments | Test methods for radiation scattered by optical components |
| Lasers and laser-related equipment | Test methods for specular reflectance and regular transmittance of optical laser components |
| Lasers and laser-related equipment | Test methods for specular reflectance and regular transmittance of optical laser components |
| Integrated optics | Interfaces — Parameters relevant to coupling properties |
| Lasers and laser-related equipment | Test methods for determination of the shape of a laser beam wavefront |
| Optics and photonics | Diffractive optics |
| Lasers and laser-related equipment | Lifetime of lasers |
| Lasers and laser-related equipment | Test methods for laser-induced damage threshold |
| Lasers and laser-related equipment | Measurement and evaluation of absorption-induced effects in laser optical components |
| Optics and photonics | Lasers and laser-related equipment — Measurement of phase retardation of optical components for polarized laser radiation |

# APPENDIX 3

# LEOMA EXECUTIVE SEMINARS

*LEOMA and the U.S. Laser Industry: The Good and Bad Moves for Trade Associations in Emerging High-Tech Industries*, First Edition. C. Breck Hitz.

| Date | Location | Speaker | Speaker | Speaker | Speaker | Speaker |
| --- | --- | --- | --- | --- | --- | --- |
| March 1990 | Monterey Hyatt Regency, Monterey, CA | Klaus Derge, VP for Europe, Spectra-Physics | James LeMunyon, DAS for Export Control, Department of Commerce | Samuel Simonsson, President, Rofin Sinar | Ed Zschau, President, Censtor (formerly U.S. Congress) | Michel Imbert, President, Optilase |
| March 1991 | Mark Hopkins Hotel, San Francisco, CA | Roger Smith, President, Silicon Valley Bank | Joseph Wahed, Chief Economist Wells Fargo Bank | Sam Colella, Partner Institutional Venture Partners | Nancy Mason, DAS for Technology Department of Commerce | James Davis, Deputy Director Lawrence Livermore |
| March 1992 | Marriott Gardens Hotel, West Palm Beach, FL | Klaus Derge, VP for Europe, Spectra-Physics | Arpad Bergh, Exec Director OIDA | Susan Lipsky, Director Strategic Partner Initiative, Department of Commerce | Tom Galantowicz, President, Newport Corporation | |
| March 1993 | Alexis Park Hotel, Las Vegas, NV | Stephen Cohen, Berkeley Roundtable on the International Economy | Walter Hoadley, Hoover Institution | James LeMunyon, President Novacon Technologies | | |
| February 1994 | Washington Marriott Hotel, Washington, DC | Duncan Moore, APS Congressional Fellow | Lionel Johns, Director White House Office of Technology Policy | Marc Stanley, Director Advanced Tech Program NIST | L.N. Durvasula, Laser Program Manager DARPA | John Mancini, Senior VP American Electronics Association |

(*Continued*)

| Date | Location | Speaker | Speaker | Speaker | Speaker | Speaker |
|---|---|---|---|---|---|---|
| March 1995 | Washington Marriott Hotel, Washington, DC | Arati Prabhaker, Director NIST | Lee Buchanan, Director Tech Reinvestment Project, Department of Commerce | Lionel Johns, Director, White House Office of Technology Policy | William Reinsch, Undersecretary for Export Controls, Dept. of Commerce | Concetto Giuliano, Director, Alliance for Photonic Tech |
| March 1996 | Washington Marriott Hotel, Washington, DC | John Bates, JAMS/Endispute | Barry Rogstad, President, American Business Conference | Stephanie Vance, legislative aide, Hon. Anna Eshoo | Alex Flint, senior staff, Hon. Pete Domenici | Cdoug Comer, Majority Staff, Director House Science Committee |
| April 1997 | DoubleTree Plaza Hotel, Washington, DC | Henry Kelly, Director Tech Policy, White House OSTP | Dorothy Dwoskin, Assistant U.S. Trade Representative | Robert Schafrick, Optics Project Manager, National Research Council | Raymond Wick, Executive Director, National Network for Electro-Optics Manufacturing | Michal Freedhoff, OSA Congressional Fellow |
| March 1998 | Latham Hotel, Washington, DC | Vince Garlock, Lead Counsel House Judiciary Committee | Keith Primdahl, U.S. Enrichment Corporation | Duncan Moore, Director Tech Policy, White House OSTP | Charles Shank, Chair Nat'l Research Council Committee on Optical Science and Engineering | James Sensenbrenner (R-WI), Chair, House Committee on Science |

# APPENDIX 4

# THE LEOMA ADR AGREEMENT

### THE LASER AND ELECTRO-OPTICS MANUFACTURERS' ASSOCIATION

#### Restated Dispute–Resolution Agreement

Whereas, the Laser and Electro-Optics Manufacturers' Association ("LEOMA") and its undersigned members desire to enter into an agreement that provides for an alternative way of resolving disputes that may arise among themselves from time to time;

Whereas, it is the intent of the parties to establish a procedure for resolving these disputes before they escalate into expensive and time-consuming lawsuits,

*Subject to the terms and conditions set forth below,* the parties agree as follows:

1. In regard to any dispute exclusively between LEOMA members who have signed this Agreement, the parties agree to make a good faith effort to settle the dispute by direct negotiations between the parties. In the event the parties are unable to settle any such dispute within thirty (30) days of the first request to negotiate the dispute, the parties

*LEOMA and the U.S. Laser Industry: The Good and Bad Moves for Trade Associations in Emerging High-Tech Industries*, First Edition. C. Breck Hitz.
© 2015 by The Institute of Electrical and Electronics Engineers, Inc. Published 2015 by John Wiley & Sons, Inc.

agree to refer the dispute to confidential mediation as set forth in Section 3 hereof.

2. This Agreement is intended to bind LEOMA members only. Parent or subsidiary corporations of LEOMA members are excluded unless they separately sign this Agreement. Likewise excluded are divisions of LEOMA members that do not meet the eligibility requirements of LEOMA.

3. The mediation process shall be commenced by one party (the "moving party") providing the other with a written request to mediate the dispute. In this notice, the moving party shall indicate which independent, professional mediator it proposes to use, and any rules and regulations that the proposed mediator suggests be used to cover the mediation. The notice will also inform the nonmoving party of the penalty for failure to respond, as set forth in Section 7 below. The nonmoving party shall respond within ten (10) days from the date of receipt of this proposal, and may either accept the proposed mediator or propose a new one as well as rules to be followed. If the parties are unable to agree on a mediator within ten (10) days of the date of the nonmoving party's response, they shall notify LEOMA offices and one shall be selected by the Executive Director of LEOMA, after consulting with both sides. If the parties are unable to agree upon rules within ten (10) days of selecting the mediator, then the mediator shall set the rules after consulting with both parties. The nonmoving party shall set the venue of the mediation process.

4. If the mediation process does not resolve the dispute, the moving party shall have the exclusive right to file a lawsuit in the jurisdiction of its choice for a period of 30 days following the end of the mediation process. Nothing herein shall limit the right of the nonmoving party to challenge the moving party's choice of jurisdiction.

5. The mediation process will continue until the dispute is resolved or until such time as the mediator and at least one of the parties finds that there is little likelihood of resolving the dispute by mediation within the next thirty (30) days. The mediator shall mediate the dispute, but shall otherwise have no power or authority to render a binding decision on the dispute. Any terms for settling the dispute shall be left up to the parties to agree upon, if possible. Each party agrees to send a representative to the mediation with full authority to settle the dispute. An appropriate non-disclosure agreement shall be signed by all parties as a condition to commencing mediation. Each party to a mediation hereunder shall bear its own costs and expenses. The expenses related to the mediator shall be divided equally among the parties.

6. No party to this Agreement may bring a lawsuit against another party to this Agreement without first engaging in the mediation process as described in Sections 2 and 3 of this Agreement, for each specific dispute that may from time to time arise between the parties, unless:

   a. the nonmoving party fails respond to the initial request for mediation within ten (10) days as set forth in Section 2 above, or

   b. the mediation process does not resolve the dispute as provided in Section 3 above, or

   c. the filing of a lawsuit is necessary in order to avoid the running of a statute of limitation that would prohibit the moving party from filing a lawsuit at a later date; but in this circumstance, the lawsuit shall not be served on the other party until such time as is the mediation process is completed or as late as possible under applicable law, or

   d. the lawsuit seeks a preliminary injunction or temporary restraining order where monetary damages would not be a sufficient remedy and a party would be irreparably harmed if it were required to comply with the provisions of this Agreement prior to obtaining a hearing on its request for injunctive relief, or

   e. ninety (90) days have expired since the date of the initial request for mediation in Section 2 above, unless both parties agree in writing to extend this date.

7. The undersigned consents to the dismissal of any lawsuit filed by the undersigned against any other party to this Agreement where the undersigned did not first comply with this Agreement unless such dismissal would result in the undersigned's claims being barred by an applicable statute of limitations. Any such dismissal shall be without prejudice.

8. A party to this Agreement may withdraw from the Agreement by providing LEOMA with written notice sixty (60) days in advance. Any disputes filed prior to or during such sixty (60) day notice period shall be governed by this Agreement, notwithstanding such notice of withdrawal. LEOMA shall publish and periodically update a list of members who have signed this Agreement and the address for sending notices hereunder and shall distribute such list to the other members who have signed the Agreement. LEOMA shall also publish and distribute to all LEOMA members notice of members who withdraw from this Agreement within ten (10) days of its receipt of the notice of withdrawal. This Agreement shall automatically terminate at such time as LEOMA withdraws or there is only one member who remains a signatory.

9. The parties acknowledge that the purpose of this Agreement is to attempt to settle disputes between LEOMA members at an early stage

and minimize the cost of litigation. Should a party to this Agreement fail to respond to a request to mediate as provided in Section 2 above, the moving party would suffer damages in terms of increased legal fees and management time spent on the dispute. Because the actual amount of damages are difficult to estimate at the present time, the parties agree that as liquidated damages a moving party shall be entitled to reimbursement of $10,000 of its legal fees in any lawsuit filed where the nonmoving party knowingly failed to respond to a request to mediate the dispute as required under Section 2.

10. This Agreement may be signed in counterparts, all of which shall be considered one agreement. It is the intent of the parties that they be bound by the terms of this Agreement with every other LEOMA member who signs this Agreement.

The undersigned is duly authorized to sign this Agreement on behalf of his or her company and has signed this Agreement as of date indicated below.

The Laser and Electro-Optics Manufacturers' Association
123 Kent Road
Pacifica, CA 94044

By:_____

Executive Director, LEOMA

Date:_____

_____

(Name of LEOMA company)

_____

_____

(Address for Notices)

By:_____

Title:_____

Date:_____

# INDEX

*LEOMA and the U.S. Laser Industry: The Good and Bad Moves for Trade Associations in Emerging High-Tech Industries*, First Edition. C. Breck Hitz.
© 2015 by The Institute of Electrical and Electronics Engineers, Inc. Published 2015 by John Wiley & Sons, Inc.

Printed in the USA
J088711SCI121214     01S29053000000000201